Springer Series in Advanced Manufacturing

Series Editor

Professor D. T. Pham
Intelligent Systems Laboratory
WDA Centre of Enterprise in Manufacturing Engineering
University of Wales Cardiff
PO Box 688
Newport Road
Cardiff
CF2 3ET
UK

Other titles published in this series

Assembly Line Design
B. Rekiek and A. Delchambre

Advances in Design
H.A. ElMaraghy and W.H. ElMaraghy (Eds.)

Effective Resource Management in Manufacturing Systems:
Optimization Algorithms in Production Planning
M. Caramia and P. Dell'Olmo

Condition Monitoring and Control for Intelligent Manufacturing
L. Wang and R.X. Gao (Eds.)

Optimal Production Planning for PCB Assembly
W. Ho and P. Ji

Trends in Supply Chain Design and Management: Technologies and Methodologies
H. Jung, F.F. Chen and B. Jeong (Eds.)

Process Planning and Scheduling for Distributed Manufacturing
Lihui Wang and Weiming Shen (Eds.)

Collaborative Product Design and Manufacturing Methodologies and Applications
W.D. Li, S.K. Ong, A.Y.C. Nee and C. McMahon (Eds.)

Decision Making in the Manufacturing Environment
R. Venkata Rao

Reverse Engineering: An Industrial Perspective
V. Raja and K. J. Fernandes (Eds.)

Yoshiaki Shimizu • Zhong Zhang
and Rafael Batres

Frontiers in Computing Technologies for Manufacturing Applications

 Springer

D9.

Yoshiaki Shimizu, Dr.Eng.
Zhong Zhang, Dr.Eng.
Rafael Batres, Dr.Eng.

Department of Production Systems
  Engineering
Toyohashi University of Technology
1-1 Hibarigaoka
Tempaku-cho
Toyohashi
Aichi 441-8580
Japan

ISBN 978-1-84628-954-5   e-ISBN 978-1-84628-955-2

Springer Series in Advanced Manufacturing ISSN 1860-5168

British Library Cataloguing in Publication Data
Shimizu, Yoshiaki
   Frontiers in computing technologies for manufacturing
   applications. - (Springer series in advanced manufacturing)
   1. Production engineering - Data processing
   I. Title II. Zhang, Zhong III. Batres, Rafael
   670.4'2'0285
ISBN-13: 9781846289545

Library of Congress Control Number: 2007931877

Printed on acid-free paper

9 8 7 6 5 4 3 2 1

springer.com

Preface

This book presents recent developments in computing technologies for manufacturing systems. It includes selected topics on information technology, data processing, algorithms and computational analysis of challenging problems found in advanced manufacturing. The book covers mainly three areas, namely advanced and combinatorial optimization, fault diagnosis, signal and image processing, and information systems. Topics related to optimization highlight on metaheuristic approaches regarding production planning, logistics network design, artificial product design, and production scheduling. The techniques presented also aim at assisting decision makers needing to consider multiple and conflicting objectives in their decision processes. In particular, this area describes the use of metaheuristic approaches to perform multi-objective optimization in terms of soft computing techniques, including the effect of parameter changes.

Fault diagnosis in manufacturing systems requires considerable experience and careful examination, which is a very time-consuming and error-prone process. To develop a diagnostic assistant computer system, methods based on cellular neural network and methods based on the wavelet transform are explained. The latter is a novel time-frequency analysis method to analyze an unsteady signal such as abnormal vibration and abnormal sound in a manufacturing system.

Topics in information systems range from web services to multi-agent applications in manufacturing. These topics will be of interest to information engineers needing practical examples for the successful integration of information in manufacturing applications.

This book is organized as follows: Chapter 1 provides a brief explanation of manufacturing systems and the roles that information technology plays in manufacturing systems. Chapter 2 focuses on several optimization methods known as metaheuristics. Hybrid approaches and robust optimization under uncertainty are also considered in this chapter. In Chap. 3, after evolutional algorithms for multi-objective analysis and solution methods associated with soft computing have been presented, the procedure of incorporating it into integrating design task is shown. The hybrid approach mentioned in the previous chapter is also extended to cover multiple objectives. Chapter 4 focuses on cellular neural networks for associative memory in intelligent sensing and diagnosis. Chapter 5 presents some useful algorithms and methods of the wavelet transform available for signal and image processing. Chapter 6 discusses methods and tools for factory and business information system integration technologies. In particular, the book includes relevant applications in every chapter to illustratively demonstrate the usage of the employed methods.

Finally, the reader will become familiar with computational technologies that can improve the performance of manufacturing systems ranging from manufacturing equipment to supply chains.

There are several ways in which this book can be utilized. It will be of interest to students in industrial engineering and mechanical engineering. The book is adequate as a supplementary text for courses dealing with multi-objective optimization in manufacturing, facility planning and simulation, sensing and fault diagnosis in manufacturing, signal and image processing for monitoring manufacturing, manufacturing systems integration, and information systems in manufacturing. It will also appeal to technical decision makers involved in production planning, logistics, supply chain and industrial ecology, manufacturing information systems, fault diagnosis, and signal processing. A variety of illustrative applications posed at the end of each chapter are intended to be useful for those professionals.

In the past decade, numerous publications have been devoted to manufacturing applications of neural networks, fuzzy logic, and evolutionary computation. Despite the large volume of publications, there are few comprehensive books addressing the applications of computational intelligence in manufacturing. In an effort to fill the void, this book has been produced to cover various topics on the manufacturing applications of computational intelligence. It contains a balanced coverage of tutorials and new results. Finally, this book is a source of new information for understanding technical details, assessing research potential, and defining future directions in the applications of computational intelligence in manufacturing.

The first idea of writing this book originated from the invitation from Mr. Anthony Doyle, Senior Editor of Engineering at the London office of the global publisher, Springer. In order to create a communication vehicle leading to advanced manufacturing, he suggested that I consider writing a book focused on the foundations and applications of tools and techniques related to decision engineering. According to this request, I asked my colleagues Zhong Zhang and Rafael Batres to join this effort by combining three primary areas of expertise.

Despite the generous assistance of so many people, some errors may still remain, for which I alone accept full responsibility.

Acknowledgments

Yoshiaki Shimizu:

I wish to express my considerable gratitude to my former colleges Jae-Kyu Yoo, now at Kanazawa University and Rei Hino now at Nagoya University for allowing me to use their collaborative works. I am indebted my students Takeshi Wada, Atsuyuki Kawada, Yasutsugu Tanaka, and Kazuki Miura for their numerical examination of the effectiveness of the methods presented in this book. I also appreciate the help of my secretary Ms. Yoshiko Nakao and my students Kanit Prasertwattana and Takashi Fujikura in word processing and drawing my awfully messy handwritten manuscript. This book would not have been completed without the continuous encouragement from my mother Toshiko, and my wife Toshika.

The help and support obtained from the publisher were also very useful. Parts of this book are based on research supported by The 21st Century COE Program "Intelligent Human Sensing," from the Japanese Ministry of Education, Culture, Sports, Science and Technology.

Zhong Zhang:

I would like to thank Yoshiaki Shimizu for inviting me to participate in the elaboration of the book. I am also grateful to Professor Hiroaki Kawabata with Okayama Prefectural University for leading me to the research of cellular neural networks and wavelet transforms. I also thank Drs. Hiroshi Toda, Michhiro Nambe and Mr. Hisanaga Fujiwaea for their collaboration in developing some of the theory and the engineering methods described in Chaps. 4 and 5. I acknowledge my students Hiroki Ikeuchi, Takuma Akiduki for implementing and refining some knowledge engineering methods. Finally, my deepest thanks go to my wife Hu Yan and my daughters Qing and Yang for their love and support.

Rafael Batres:

I would like to thank Yoshiaki Shimizu for inviting me to participate in the elaboration of the book since its original concept. I am also grateful to Yuji Naka for planting the seed that gave me a holistic understanding of systems thinking. Special thanks are due to Matthew West for his countless useful discussions on the ISO 15926 upper ontology. I also thank David Leal and David Price for their collaboration in developing the OWL version of the ontology. I would like to give recognition to Steven Kraines (University of Tokyo) and Vincent Wolowski for letting me participate in the development of cognitive agents. I acknowledge my students Masaki Katsube, Takashi Suzuki, Yoh Azuma and Mikiya Suzuki for implementing and refining some of the knowledge engineering methods described in Chap. 6. On the personal level, I would like to thank my wife Haixia and my children Joshua and Abraham for their support, love and patience.

Toyohashi   Yoshiaki Shimizu
March 2007   Zhong Zhang
                                                Rafael Batres
                          Toyohashi University of Technology

Contents

1   **Introduction** 1
   1.1 Manufacturing Systems 1
   1.2 The Manufacturing Process 3
   1.3 Computing Technologies 4
   1.4 About This Book 9
   References 11

2   **Metaheuristic Optimization in Certain and Uncertain**
   Environments 13
   2.1 Introduction 13
   2.2 Metaheuristic Approaches to Optimization 13
      2.2.1 Genetic Algorithms 14
      2.2.2 Simulated Annealing 22
      2.2.3 Tabu Search 26
      2.2.4 Differential Evolution (DE) 27
      2.2.5 Particle Swarm Optimization (PSO) 32
      2.2.6 Other Methods 34
   2.3 Hybrid Approaches to Optimization 36
   2.4 Applications for Manufacturing Planning and Operation 38
      2.4.1 Logistic Optimization Using Hybrid Tabu Search 39
      2.4.2 Sequencing Planning for a Mixed-model Assembly
           Line Using SA 48
      2.4.3 General Scheduling Considering Human–Machine
           Cooperation 53
   2.5 Optimization under Uncertainty 60
      2.5.1 A GA to Derive an Insensitive Solution against
           Uncertain Parameters 60
      2.5.2 Flexible Logistic Network Design Optimization 65
   2.6 Chapter Summary 71
   References 72

3   Multi-objective Optimization Through Soft Computing Approaches 77
  3.1   Introduction 77
  3.2   Multi-objective Metaheuristic Methods 79
        3.2.1   Aggregating Function Approaches 80
        3.2.2   Population-oriented Approaches 80
        3.2.3   Pareto-based Approaches 82
  3.3   Multi-objective Optimization in Terms of Soft Computing 87
        3.3.1   Value Function Modeling Using Artificial Neural
                Networks 88
        3.3.2   Hybrid GA for Solving MIP under Multi-objectives 91
        3.3.3   MOON^{2R} and MOON^2 95
  3.4   Applications of MOSC for Manufacturing Optimization 105
        3.4.1   Multi-objective Site Location of Waste Disposal
                Facilities 106
        3.4.2   Multi-objective Scheduling of Flow Shop 108
        3.4.3   Artificial Product Design 112
  3.5   Chapter Summary 121
  References 122

4   Cellular Neural Networks in Intelligent Sensing and Diagnosis 125
  4.1   The Cellular Neural Network as an Associative Memory 125
  4.2   Design Method of CNN 128
        4.2.1   A Method Using Singular Value Decomposition 128
        4.2.2   Multi-output Function Design 131
        4.2.3   Un-uniform Neighborhood 135
        4.2.4   Multi-memory Tables for CNN 140
  4.3   Applications in Intelligent Sensing and Diagnosis 143
        4.3.1   Liver Disease Diagnosis 143
        4.3.2   Abnormal Car Sound Detection 147
        4.3.3   Pattern Classification 152
  4.4   Chapter Summary 155
  References 156

5   The Wavelet Transform in Signal and Image Processing ... 159
  5.1   Introduction to Wavelet Transforms 159
  5.2   The Continuous Wavelet Transform 160
        5.2.1   The Conventional Continuous Wavelet Transform 160
        5.2.2   The New Wavelet: The RI-Spline Wavelet 162
        5.2.3   Fast Algorithms in the Frequency Domain 167
        5.2.4   Creating a Novel Real Signal Mother Wavelet 173
  5.3   Translation Invariance Complex Discrete Wavelet Transforms . 180
        5.3.1   Traditional Discrete Wavelet Transforms 180

5.3.2   RI-spline Wavelet for Complex Discrete Wavelet
Transforms 182
5.3.3   Coherent Dual-tree Algorithm 185
5.3.4   2-D Complex Discrete Wavelet Transforms 189
5.4   Applications in Signal and Image Processing 194
5.4.1   Fractal Analysis Using the Fast Continuous Wavelet
Transform 194
5.4.2   Knocking Detection Using Wavelet Instantaneous
Correlation 200
5.4.3   De-noising by Complex Discrete Wavelet Transforms ... 205
5.4.4   Image Processing and Direction Selection 212
5.5   Chapter Summary 217
References 219

6   Integration of Information Systems 221
6.1   Introduction 221
6.2   Enterprise Systems 224
6.3   MES Systems 224
6.4   Integration Layers 225
6.5   Integration Technologies 225
6.5.1   Database Integration 225
6.5.2   Remote Procedure Calls 226
6.5.3   OPC 227
6.5.4   Publish and Subscribe 227
6.5.5   Web Services 228
6.6   Multi-agent Systems 229
6.6.1   FIPA: A Standard for Agent Systems 230
6.7   Applications of Multi-agent Systems in Manufacturing 232
6.7.1   Multi-agent System Example 232
6.8   Standard Reference Models 236
6.8.1   ISO TC184 236
6.9   IEC/ISO 62264 237
6.10 Formal Languages 240
6.10.1 EXPRESS 240
6.10.2 Ontology Languages 240
6.10.3 OWL 241
6.10.4 Matchmaking Agents Revisited 242
6.11 Upper Ontologies 243
6.11.1 ISO 15926 244
6.11.2 Connectivity and Composition 244
6.11.3 Physical Quantities 246
6.12 Time-reasoning 249
6.13 Chapter Summary 250
References 251

7   Summary 253

A   Introduction to IDEF0 259
     References 261

B   The Basis of Optimization Under a Single Objective 263
     B.1   Introduction 263
     B.2   Linear Programming and Some Remarks on Its Advances 264
     B.3   Non-linear Programs 269
     References 275

C   The Basis of Optimization Under Multiple Objectives 277
     C.1   Binary Relations and Preference Order 277
     C.2   Traditional Methods 279
          C.2.1   Multi-objective Analysis......................... 279
          C.2.2   Prior Articulation Methods of MOP 281
          C.2.3   Some Interactive Methods of MOP 283
     C.3   Worth Assessment and the Analytic Hierarchical Process 290
          C.3.1   Worth Assessment 290
          C.3.2   The Analytic Hierarchy Process (AHP) 291
     References 294

D   The Basis of Neural Networks 297
     D.1   The Back Propagation Network 297
     D.2   The Radial-basis Function Network 299
     References 301

E   The Level Partition Algorithm of ISM 303
     References 305

Index 307

1

Introduction

1.1 Manufacturing Systems

The etymology of the word manufacturing stems from of the Latin word "manus", which means hand and the Latin word "factura" which is the past participle of "facere" meaning "made". It thus refers to a "making" activity carried out by hand, which can be traced back to ancient times when the "homo faber", the toolmaker, invented tools and implements in order to survive [1]. The evolution of manufacturing systems is shown in Figure 1.1.

An enterprise implements a manufacturing system that uses resources such as energy, materials, currency, labor, machines and knowledge to produce value-added products (new materials, assembled products, energy or services).

Earlier attempts to understand the nature of manufacturing systems viewed production processes as an assembly of parts each dedicated to one specific function. For example, Taylor who introduced the concept of "scientific management" perceived tasks, equipment, and labor as interchangeable and passive parts. In order to increase production and quality, each production task had to be analyzed in terms of its basic elements to develop specialized equipment and labor to attain their optimal performance. In other words, organizations that implemented Taylor's ideas devised ways to optimize parts individually, which often resulted in a suboptimal performance of the whole system: the whole was the sum of the parts.

A group of researchers firstly challenged this view during the 1940s. This multi-disciplinary group of scientists and engineers introduced the notion of systems thinking, which is described in the work of Bertalanffy [2], Ackoff [3], and Checkland [4]. Contrasting with Taylor's approach, systems thinking is based on the assumption that the performance of the whole is affected by a synergistic interaction of its parts. In other words, the whole is more than the sum of the parts, which implies that there are some emergent properties of the whole that cannot be explained by looking at the parts individually. Consequently, it became possible to develop complex models to describe the behavior of materials and machines, which had an enormous impact in almost

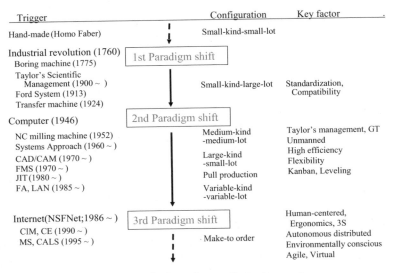

Fig. 1.1. Evolution of manufacturing systems

all areas of manufacturing to the extent that today's factories and products are unthinkable without such models. Subsequently, as noted by Bertalanffy, systems were conceived as open structures that interact with their surroundings. This in turn led researchers and practitioners to realize that production systems should not be viewed independently from societal and environmental systems.

With the advent of the computer, it became possible to analyze models, carry out optimization and solve other complex mathematical problems that had been difficult or impossible to cope with. Subsequently, the systems approach gave rise to a number of new fields such as control theory, computer science, information theory, automata theory, artificial intelligence, and computer modeling and simulation science [5]. Many concepts central to these fields have found practical applications in manufacturing, including neural networks (NN), Kalman filters, cellular automata, feedback control, fuzzy logic, Markov chains, evolutionary algorithms (EA), game theory, and decision theory [6]. Some of these concepts and their applications are discussed in this book.

Along with the development of such a systems approach, the mass data processing ability of the computer has enabled manufacturing systems to produce more diversely and more efficiently. Numerically controlled machinery like CNC (computerized numerical control), AGV (automated guided vehicle) and industrial robots were invented, and automation and unmanned production became possible in the 1980s.

Manufacturing systems were originally centered on the factory. However, social and market forces have compelled industries to extend the system boundaries to develop high-quality products in shorter time and at less cost. Nowadays, manufacturing systems can encompass whole value chains involving raw material production, product manufacturing, delivery to final consumers, and recycling of materials.

In addition to the computer-aided technologies like CAD/CAM, CAP, *etc.*, ideas of organization and integration of individual systems are incorporated in FMS (flexible manufacturing system), FA (factory automation) and CIM (computer integrated manufacturing).

The third paradigm shift brought about by information technology (IT) has been accelerating current agile manufacturing increasingly. IT plays an essential part the realization of the emerging systems and technologies like IMS (intelligent manufacturing system), CALS (computer-aided logistic support/ commerce at light speed), CE (concurrent engineering), *etc.* They are the fruits of computational intelligence [7], software integration, collaboration and autonomous distributed concepts via an information network. The more sophisticated development from those factors must be directed towards the sustainable progress of manufacturing systems [8] so that difficulties left unsolved will be removed from in the next generation. A road map of the forthcoming manufacturing system should be substantially drawn to consider 3S, *i.e.*, customer satisfaction, employee satisfaction and social satisfaction, while making an earnest effort to attain environmentally conscious manufacturing and human-centered manufacturing.

1.2 The Manufacturing Process

A basic structure as a transformation process in manufacturing system is depicted in Figure 1.2 in terms of the IDEF0 modeling technique ([1], see to Appendix A). It can describe suitably not only what is done by the process, but also why and how it is done, associated with major three basic elements of manufacturing, namely, object (input/ output), mean (mechanism) , and constraint (control). Inputs represent things to be changed by the process into outputs. The mechanisms refer to actors, or instrument resources necessary to carry out the process, such as machineries, tools, databases, and personnel. The control or constraint for a manufacturing process correspond to production requirements, production plans, production recipes, and so on.

From a different viewpoint [10], we can see the manufacturing system as a reality filling a structural function that concerns space layout and contributes to increase the efficiency of the flow of material. In addition, its procedural function is embedded in a series of phases in manufacturing system (see Figure 1.3) to achieve the ultimate goal. This involves strategic planning such as project planning, which inter-relates with the outer world of a manufacturing

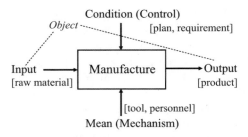

Fig. 1.2. Basic structure of manufacturing

system or market. Tactical planning serves the inside part of the manufacturing system, and is classified into long-term planning, medium-term planning, and short-term planning or production scheduling. Also operation and production control have many links with the procedural function of manufacturing.

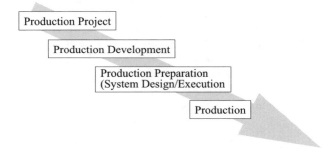

Fig. 1.3. Procedure phase in manufacturing system

In current engineering, since configuration of such a manufacturing process is not only large but also complex and complicated, the role of information system in managing the whole system consistently becomes extremely important. For example, from raw material procurement to product delivery, information systems have become ubiquitous assets in the supply chain. Information visibility has become a key factor that can significantly influence the decision making in the supply chain by allowing shorter lead times, and reducing inventories and transportation costs.

1.3 Computing Technologies

Nowadays, the interdisciplinary environment of research and development is truly required to deal with the complexity related to systems such as biology and medicine, humanities, management sciences and social sciences. Intelligence for computing technologies is creativity for analyzing, designing, and developing intelligent systems based on science and technology. This has opened new dimensions for scientific discovery and offered a scientific breakthrough. Consequently, applications of computing technologies in manufacturing will play a leading role in the development of intelligent manufacturing systems whose wide spectrum include system design and planning, process monitoring, process control, product quality control, and equipment fault diagnosis [11].

From such viewpoints, this book is concerned with recent advances in methodologies and applications of metaheuristics, soft computing (SC), signal processing, and information technologies. Thus, the book covers topics such as combinatorial and multi-objective optimizations (MOP), neural networks, wavelet and information technologies like intelligent agents and ontologies.

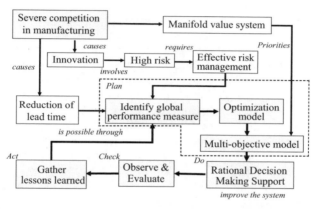

Fig. 1.4. Root-cause analysis toward rational decision-making

The interest in optimization is due to the fact that companies are looking for ways to improve the manufacturing system and reduce the lead time. In order to improve a manufacturing system, it becomes necessary to identify global performance parameters, which is possible through root-cause analysis techniques such as in the PDCA cycle. A PDCA cycle is a generic methodology for continuous improvement that is based on the "plan, do, check, act" cycle borrowed from the total quality management philosophy introduced to Japan by W. Edwards Deming [12]. The PDCA cycle, which is also known as Shewhart cycle (named after Deming's teacher Walter A. Shewhart) comprises four steps:

1. Study a system to decide what change might improve it (plan)
2. Carry out tests or modify the system (do)
3. Observe and evaluate the effect (check)
4. Gather lessons learned (act)

Once the global performance measures are identified in Step 1, the system is modeled and the optimum values for the performance measures are obtained which is the foundation for Step 2. This brings us to the first class of computing technologies, which covers methods and tools dealing with how to obtain the optimum values of the performance measures (see Figure 1.4). Special emphasis is placed on metaheuristic methods, multi-objective optimization, and soft computing.

The term metaheuristic is composed of the Greek prefix "meta" (beyond) and "heuriskein" (to find), and represents the generic name of every heuristic method, including evolutionary algorithms . An approximated solution with good quality is shown to be obtained within an acceptable computation time through a variety of applications. Roughly speaking, they require no mathematically rigid procedures and aim at attaining the global optimum. In addition, most commonly used methods are targeted at combinatorial optimization problems that have great potential applications in recent manufacturing systems. These are special advantages concerned with real world problems for which there has been no satisfactory algorithm. They also have the potential of coping with uncertainties involved in the mathematical formulation in a rational way. It is of special importance to present a flexible and/or robust solution for uncertainties.

The need for agile and flexible manufacturing is accelerated under the diversified customer demands. Under such circumstances, it is often adequate to formulate the optimization problem as one in which there are several criteria or objectives. Usually, since such objectives involve some that conflict with each other, the articulation among them becomes necessary to find the best compromise solution. This type of problem is known as either a multi-objective, multi-criteria, or a vector optimization problem. Multi-objective optimization is a powerful tool available for manifold and flexible decision-making in manufacturing systems.

On the other hand, soft computing (SC) is a collection of new computational techniques in computer science, artificial intelligence, and machine learning. The basic ideas underlying SC is largely due to earlier studies on fuzzy set theory by Zadeh [13, 14]

The most important areas of soft computing are as follows:

1. Neural networks (NN)
2. Fuzzy systems (FS)
3. Evolutionary computation (EC) including evolutionary algorithms and swarm intelligence
4. Ideas on probability including the Bayesian network and chaos theory

SC differs from conventional (hard) computing mainly in two aspects: it is more tolerant of imprecision, uncertainty, partial truth, and approximation; it weight inductive reasoning more heavily. Moreover, since SC techniques [15, 16] are often used to complement each other in applications, new hybrid approaches are expected to be invented by a particularly effective combination ("neuro–fuzzy systems" is a striking example). The multi-objective optimization method mentioned in Chap. 3 presents a new type of partnership in which each partner contributes a distinct methodology for addressing problems in its domain. Such an approach is likely to play an especially important role and, in many ways, facilitate a significant paradigm shift of computing technologies targeting manufacturing systems.

To diagnose manufacturing systems, engineers must base their judgments on tests and much measurement data. This requires considerable experience and careful examination, which is a very time-consuming and error-prone process. It would be desirable to develop a computer diagnostic assistant based on the knowledge of technological specialists, which may improve the correct diagnosis rate. However, unsteady fluctuations in the first problem samples make it very difficult to develop a reliable diagnosis system.

Humans have a spectacular capability of processing ambiguous information very skillfully. The artificial neural network is a kind of information processing system made by modeling the brain on a computer and has been developed to realize this peculiar human capability. Typical models of neural networks are multi-layered models such as the conventional perceptron-based neural networks (NN) that have been applied to machine learning. They have the structure of a black box system and can reveal the incorrect recognition. On the other hand, Hopfield neural networks are cross-coupled attractor models that incorporate existing knowledge to investigate the reason for incorrect recognition.

Furthermore, the cellular neural network (CNN) [17] as a cross-coupled attractor models has called for special attention due to the possibility of wide applications. Recently its concrete design method for associative memory has been proposed [18]. Since then, some further applications have been proposed, but studies on improving its capability are few. Some researchers have already shown CNN to be effective for image processing. Hence, if the advanced association CNN system has been provided, the CNN recognition system will be established in manufacturing system.

As is well known, signal analysis and image processing are very important in manufacturing systems. A signal can be generally divided into a steady signal and an unsteady signal. Many signals such as abnormal vibration, and abnormal sound can be considered as unsteady signals. An important characteristic of the unsteady signal is that each frequency component changes with time. To analyze an unsteady signal, therefore, we need a time-frequency analysis method. Accordingly, some standard methods have been proposed and applied in various research fields.

The Wigner distribution (joint time-frequency analysis) and the short time Fourier transform are typical. However, when the signal includes two or more characteristic frequencies, the Wigner distribution suffers from the contamination referring to the cross terms. That is, the Wigner distribution can yield imperfect information about the distribution of energy in the time-frequency plane. The short time Fourier transform is probably the most common approach for analyzing unsteady signals of sound and vibration. It subdivides the signal into short time segments (this is same as using a small window to divide the signal), and apply a discrete Fourier transform to each of these. However, since the window whose length may vary with each frequency component is fixed, it is unable to obtain optimal results for individual frequency components. On the other hand, the wavelet transform, which is a time-frequency methods, does not have such problems and has some desirable properties for unsteady signal analysis and applications in various fields.

Motivated with more effective decision support on production, information systems were first introduced on the factory floor and the tendency to automation continues today. For example, the use of real-time data allows for better scheduling and maintenance. With such information available, manufacturers have realized that they can use equipment and other resources more efficiently. Additionally, timely decisions and more rational planning translate into reduction of wear-and-tear on equipment. Used as stand-alone applications, plant information systems provide enough valuable information to justify their use. However, information systems seen from a wider perspective can only serve this purpose when there are sufficient linkages between the individual information systems within the manufacturing system.

On the other hand, investments in information technology tend to increase to the extent that the advantages are overshadowed by the incurred costs. World-wide enterprises are spending up to 40% of their IT budget on data integration. For manufacturing companies this budget reflects the phenomena of rapidly changing technologies, and the difficulties in integrating software from different vendors and legacy systems. A single stake-holder in the supply chain may have as much as 150 different applications where attempts to integrate them can be up to five times the cost of the application software. This may explain the increase in the demand of system integration professionals during the last decade. A variety of technologies have been developed that facilitates the task of integrating different applications. However, this situation demands system integrators to be proficient in many, if not in all, of applications. Furthermore, integration technologies tend to evolve very quickly. Current integration technologies and ongoing research in this area are discussed in further detail in the rest of the chapter.

An even more difficult challenge is not in the connectivity between systems themselves but lies in the meaning of the data. Putting this differently, the same word can have different meanings in different applications. For example, the term resource as used in one application may refer to equipment alone, while the same term in another application may mean equipment, personnel or material. In fact, one of the authors is aware of a scheduling tool in which the term resource is used to represent both equipment and personnel! To solve the problem of the meaning of information, several standardization activities are being carried out world-wide, ranging from batch information systems to enterprise resource planning systems. Many successful integration projects have become possible through the implementation of such standards. However, with current database technologies, information engineers tend to focus on data rather than on what exists in reality. This can lead to costly updates of the information models as technology evolves. Knowledge engineering specialists in industry and academia have already started to address this problem by developing ontologies and tools. An ontology is a theory of reality that "describe the kinds and structures of objects, properties, events, processes, and relations in every area of reality", which allows dynamic integration of information that cannot be achieved with conventional database systems. Specific applications of ontologies in the manufacturing domain are explained in detail in Chap. 6.

1.4 About This Book

This book presents an overview of the state of the art of intelligent computing in manufacturing and presents the selected topics on modeling, data processing, algorithms, and computational analysis for intelligent manufacturing systems. It introduces the various approaches to dealing with difficult problems found in advanced manufacturing. It includes three big areas, which are not taken into account elsewhere together in a consistent manner, namely combinatorial and multi-objective optimizations, fault diagnosis and monitoring, and information systems. The techniques presented in the book aim at assisting decision makers needing to consider multiple, conflicting objectives in their decision processes and should be of interest to information engineers needing practical examples for the successful signal processing and sensing, and integration of information in manufacturing applications.

The book is organized as depicted in Figure 1.5 where four keywords extracted from the title are deployed. Chapter 1 provides a brief explanation of manufacturing systems and our viewpoints in order to explain the developments in the emerging manufacturing systems.

Chapter 2 focuses on several optimization methods known as metaheuristics. They are particularly effective for dealing with combinational optimization problems that are becoming very important for various types of problem-solving in manufacturing. Hybrid approaches and robust optimization under uncertainty associated with metaheuristics are also considered in this chapter.

In Chap. 3, after the introduction of evolutional algorithms for multi-objective analysis, a new discovery of multi-objective optimization is presented to show the solution method associated with soft computing and the procedure

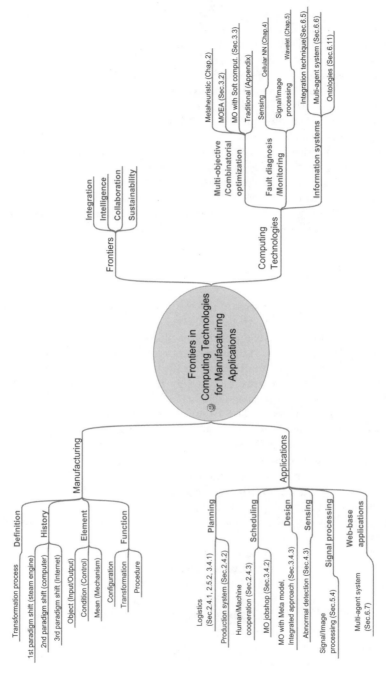

Fig. 1.5. A glance at book contents

integrating it into the design task. The hybrid approach mentioned in the foregoing chapter is also extended under multiple objectives.

Chapter 4 focuses on CNN for associative memory and explains common design methods by using a singular value decomposition. After some new models such as the multi-valued output CNN and the multi-memory tables CNN are introduced, they are applied to intelligent sensing and diagnosis. The results in this chapter contribute to improving the capability of CNN for associative memory and the future possibility as the memory medium.

In Chap. 5, by taking the wavelet transform, some useful algorithms and methods are shown such as a fast algorithm in the frequency domain for continuous wavelet transform, a wavelet instantaneous correlation method by using the real signal mother wavelet, and a complex discrete wavelet transform through the real-imaginary spline wavelet.

Chapter 6 discusses methods and tools for factory and business information systems. Some of the most common integration technologies are discussed. Also, new techniques and methodologies are presented.

In particular, the book presents the relevant applications in each chapter to illustratively demonstrate usage of the employed methods. A number of appendices are given for the sake of convenience. As well as supplementing the explanation in the main text, a few of the appendices aim to fuse traditional knowledge with recent knowledge, and to facilitate the generation of new meta-ideas by borrowing some from the old.

The aim of this book is to present the state of the art and highlight the recent advances both of methodologies and applications of computing technologies in manufacturing. We hope that this book will help the reader to develop insights for creating and managing manufacturing systems that improve people's life while making a sustainable use of the resources of this planet.

References

1. Arendt H (1958) The human condition. University of Chicago Press, Chicago
2. Bertalanffy L (1976) General system theory. George Braziller, New York
3. Ackoff RL (1962) Scientific methods: optimizing applied research decisions. Wiley, New York
4. Checkland P (1999) Systems thinking, systems practice. Wiley, New York
5. Heylighen F, Joslyn C, Meyers RA (eds.) (2001) Encyclopedia of physical science and technology (3rd ed.). Academic Press, New York
6. Schwaninger M (2006) System dynamics and the evolution of the systems movement, systems research and behavioral science. System Research, 23:583–594
7. Kusiak A (2000) Computational intelligence in design and manufacturing. Wiley, New York
8. Graedel T E, Allenby B R (1995) Industrial ecology. Prentice Hall, Englewood Cliffs, NJ
9. Marca DA, McGowan CL (1993) IDEF0/SADT business process and enterprise modeling. Eclectic Solutions Corporation, San Diego, CA

10. Hitomi K (1996) Manufacturing systems engineering (2nd ed.). Taylor & Francis, London
11. Wang J, Kusiak A (eds.) (2001) Computational intelligence in manufacturing handbook. CRC Press, Boca Raton
12. Cornesky B (1994) Using the PDCA model effectively. TQM in Higher Education, August, 5
13. Zadeh LA (1965) Fuzzy sets. Information and Control, 8:338–353
14. Zadeh LA, Fu K-S, Tanaka K, Shimura M (eds.) (1975) Fuzzy sets and their applications to cognitive and decision processes. Academic Press, London
15. Suzuki Y, Ovaska S, Furuhashi T, Roy R, Dote Y (eds.) (2000) Soft computing in industrial applications. Springer, London
16. Kecman V (2001) Learning and soft computing: support vector machines, neural networks, and fuzzy logic models. A Bradford Book, MIT Press, Cambridge
17. Chua L O, Yang L (1988) Cellular neural networks: theory. IEEE Transaction of Circuits and System, CAS-3:1257–1272
18. Liu D, Michel AN (1993) Cellular neural networks for associative memories. IEEE Transaction of Circuits and Systems, CAS-40:119–121

Metaheuristic Optimization in Certain and Uncertain Environments

2.1 Introduction

Until now, a variety of optimization methods have been used as effective tools for making a rational decision in manufacturing systems and will surely continue to do so. By virtue of the outstanding progress in computers, many applications have been carried out in the real world using commercial software that has been developed greatly. To understand the proper usage of software and the adequate choice of optimization method through revealing merits and demerits compared with recent metaheuristic approaches, it is essential for every practician to have basic knowledge of these methods.

We can always systematically define every optimization problem by the triplet of arguments $(x, f(x), X)$ where x is an n-dimensional vector called decision variable and $f(x)$ an objective function. Moreover, X denotes a subset of R^n called an admissible region or a feasible region that is prescribed generally by a set of equality and/or inequality equations called constraints. Using these arguments, the optimization problem can be described generally and simply as follows:

$$[Problem] \quad \min \quad f(x) \text{ subject to } x \in X.$$

The maximization problem can be handled in the same way as the minimization problem just by multiplying the objective function by -1. By combining different properties of each arguments of the triplet, we can define a variety of optimization problems. A brief introduction to the traditional optimization method is given in Appendix B.

2.2 Metaheuristic Approaches to Optimization

In this section, we will review several emerging methods known as metaheuristic optimizations. Roughly speaking, metaheuristic optimizations are considered as a kind of direct search method aiming at a global optimum by utilizing a certain probabilistic drift and heuristic idea. The algorithms are commonly depicted as shown in Figure 2.1. To give a certain perturbation to the current (tentative) solution, a candidate solution will be generated. It is in turn evaluated through comparison with the tentative solution. Not only when the candidate is superior to the tentative (downhill move), but also when it is a bit inferior (uphill move), the candidate solution can become a new tentative solution with the prescribed probability. By occasionally accepting an inferior candidate (uphill more), these methods can escape from the local optimum and attain the global optimum as illustrated in Figure 2.2.

From these tactics, the algorithms are mainly characterized by the manners in which to derive the tentative, how to nominate the candidate, and how to decide the solution update. Metaheuristic optimization can also readily cope with even the combinatorial optimization. Due to these favorable properties and support by the outstanding progress both of computer hardware and software, these methods have been widely applied to solve difficult problems in recent manufacturing optimization [1, 2].

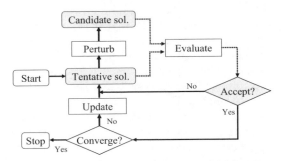

Fig. 2.1. General procedure of the metaheuristic approach

2.2.1 Genetic Algorithms

Genetic algorithm (GA) [3, 4, 13] is a pioneering method of metaheuristic optimization which originated from the studies of cellular automata of Holland [6] in the 1970s. It is also known as an evolutionary algorithm and a search technique that copies from biological evolution. In GA, a population of candidate solutions called individuals evolves toward better solutions from generation to generation. Since it needs no difficult mathematical conditions and can perform well with all types of optimization problems, it has been widely applied to solve problems not only in the engineering field but also in art, biology, economics, operations research, robotics, social sciences, and so on. The algorithm is closely related to some terminologies of natural selection,

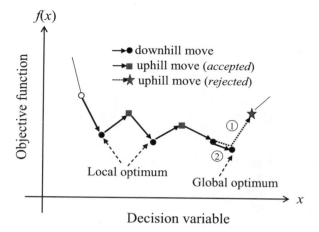

Fig. 2.2. Escape from the local search in terms of probabilistic drift

e.g., population, generation, fitness, *etc.*, and is composed of genetic operators such as reproduction, mutation and recombination or crossover.

Below, a typical algorithm of GA is described by illustration for the unconstrained optimization problem, *i.e.*, minimize $f(x)$ with respect to $x \in \mathrm{R}^n$. An n-dimensional decision variable x or solution is corresponded to a chromosome or individual that is a string of genes, and its value is represented by appropriate notations depending on the problem. The simplest algorithm represents each chromosome as a bit string. Other variants treat the chromosome as a list of numbers, nodes in a linked list, hashes, objects, or any other imaginable data structure. This procedure is known as coding, and is described as follows assuming, for simplicity, the decision variable is scalar:

$$x := G_1 \cdot G_2 \cdots G_i \cdots G_L,$$

where G_i denotes the gene, L length of the string, and the position in the string is called locus. An allele is a kind of gene and takes 0 or 1 in the simplest binary representation. This representation is called a genotype. After the evolution in the procedure, the genotype is returned to the value (phenotype) through the reverse procedure of encoding (decoding) for evaluating the objective function numerically. Usually, the length of chromosome is fixed, but variable representations are also used (in this case, the crossover implementation mentioned below becomes more complex).

The evolution is started by randomly generating individuals, each of which corresponds to a solution. A set of individuals is called a population (population-based algorithm). Traditionally, the initial population is generated to cover the entire search space. During each successive generation, a new population is stochastically selected from the current population based on its fitness. Contrasting the iteration with the generation, the optimization process is defined by the search on a solution set $P_{OP}(t)$ described at the t-th generation as follows:

$$P_{OP}(t) = \{x_{1,t}, x_{2,t}, \ldots, x_{N_p,t}, \} \tag{2.1}$$

where N_p denotes a population size, and $x_{i,t}$, $(i = 1, 2, \ldots, N_p)$ is supposed to be a genotype. When we do need to note the generation explicitly, x_i means $x_{i,t}$ hereinafter.

At each generation, the fitness of whole population is evaluated, and the survival individuals are selected through a reproduction process where fitter solutions are more likely to be selected. Simply, the objective function is amenable to the fitness function of x_i, i.e., $F_i = f(x_i), (\geq 0)$. To keep regularity and increase the efficiency, however, the original value should be transformed into the more proper value using a certain scaling technique. The following are typical scaling methods:

1. linear scaling
2. sigma truncation
3. power law scaling

In the above, linear scaling simply applies a linear transformation to F_i (≥ 0)

$$\hat{F}_i = aF_i + b,$$

where a and b are appropriately chosen coefficients. Sigma truncation is applied as

$$\hat{F}_i = aF_i - (\bar{F} - c\sigma),$$

where $\bar{F}$ and σ denote the average of the fitness over the population and its standard deviation, respectively. Moreover, c is a parameter between $1 \sim 3$. Finally, power law scaling is described as

$$\hat{F}_i = (F_i)^k, \ (k > 1).$$

Since the implementation and the evaluation of the fitness are important factors affecting the speed and efficiency of the algorithm, the scaling has a particular significance. Evolution or search takes place through genetic operators such as reproduction, mutation and crossover, each of which will be explained below.

A. Rule of Reproduction

As to why the rule of natural selection is applied to the optimization may rely on an observation that the better solutions often locate in the niche of good solutions found so far. This is compared to a concept regarding the stationary condition for optimization in a mathematical sense. The following rules are popularly known as the reproduction:

1. Roulette selection: This applies the rule that individuals can survive into the next generation based on the rate of fitness value of each (F_i) to the total value ($F_T = \sum_{k=1}^{N_p} F_k$), *i.e.*, $p_i = F_i/F_T$, as shown in Figure 2.3. This can constitute a rationale such that an individual with a greater fitness has a larger possibility of being selected in the next generation; an individual with even a low fitness has a chance of being selected. For these reasons, we can maintain the manifold of the population, and prevent it from being trapped at the local optimum. In addition, since this rule is simple, it is considered as a basic rule in the reproduction of GAs. There are two variants of this rule.

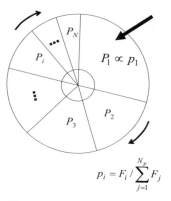

$$p_i = F_i / \sum_{j=1}^{N_p} F_j$$

Fig. 2.3. Roulette selection

- Proportion selection: Generating a random value between [0,1] (denoted by rand()), search the minimum k satisfying the condition such that $\sum_{i=1}^{k} F_i \geq \text{rand}() F_T$. Then the k-th individual can survive. This procedure is repeated until the total number of survivors becomes N_p.
- Expected-value selection: The above methods sometimes cause an undue selection due to probabilistic drift when the number of population is not sufficiently large. To fix this problem, this method tries to select the individual in terms of the expected value based on the rate p_i. That is, when the required number of selections is N_s, the i-th individual can propagate by $[p_i N_s]$. Here $[\cdot]$ denotes the Gauss symbol.

2. Ranking selection: This method can fix a certain problem occurring with roulette selection. Let us consider a situation where there exist individuals with extremely high fitness values, or there is almost no difference among the fitness values of individuals. In the former case, it can happen that only the particular individuals will flourish, while in the latter every individual dwells on the average and the better ones cannot grow for ever. Instead of the magnitude of fitness itself, it is possible to achieve a proportional selection by paying attention to the ranking. According to the magnitude

of fitness, rank the individual first. Then the selection will take place in the order of the selection rate decided *apriori*. For example, linear ranking sets up the selection rate for the individual at the i-th place of the ranking as $p_i = a - b(i - 1)$ meanwhile non-linear one as $p_i = c(1 - c)^{i-1}$, where a, b, and c are coefficients in $(0, 1)$.

3. Tournament selection: In this method, the individuals with the highest fitness among the fixed size of the sub-population selected randomly will survive through tournament. This procedure is repeated until the predetermined number of selections has been attained.

4. Elitist preserving selection: If we select by relying only on a probabilistic basis, favorable individuals happen to disappear due to the probability drift also imbedded in genetic operations like crossover and mutation. This phenomenon may cause a performance degradation known as premature convergence. Though this is the generic nature of GA, it has a side effect of preventing trapping at the local optimum. Noticing these facts, this selection preserves the elite in the present population without any reserve for the next generation. This has a certain effect of preventing that the best be killed through the genetic operations, but, in turn, produces the risk of another convergence. Consequently, this method should be applied together with another selection method. Obviously, under this rule the highest value of fitness increases monotonically along with the generation.

B. Crossover

This operation plays the most important role in GA. Through the crossover, a pair selected randomly from the population becomes parents, and produce a pair of offspring that share the characteristics of their parents by exchanging genes with each other. To use this mechanism, we need to define properly three routines: how to select the pairs, how to recombine the chromosome, and how to migrate the offspring into the population. Though various crossover methods have been proposed, depending on the problem, below we show only a few typical methods for the case of binary coding, *i.e.*, $\{0, 1\}$, for simplicity.

1. One-point crossover
   Select randomly a crossover point in the string of parents and exchange the right-hand parts mutually. (Below "|" represents the crossover point)

Parent 1 : 01001|101	Offspring 1 : 01001|110
Parent 2 : 01100|110	Offspring 2 : 01100|101

2. Multi-point crossover
   Select randomly plural crossover points in the string and exchange the parts mutually. (See below for the two-point crossover)

Parent 1 : 010|011|01	Offspring 1 : 010|001|01
Parent 2 : 011|001|10	Offspring 2 : 011|011|10

3. Uniform crossover

This method first prepares a mask pattern by generating {0, 1} uniformly at every locus beforehand. Then offspring "1" inherits the character of parent "1" if the allele of the mask pattern is 1, and parent "2" if it is 0. Meanwhile, offspring "2" is generated in an opposite manner. See the following example, which assumes that the mask pattern is given as 01101101:

$$\text{Parent 1}:\quad 01001101 \qquad \text{Offspring 1}:\quad 01001111$$
$$\text{Parent 2}:\quad 01100110 \qquad \text{Offspring 2}:\quad 01100100$$

The simple crossover operates as follows:

Step 1: Set $k = 1$.

Step 2: Select randomly a pair of individuals (parents) from among the population.

Step 3: Apply an appropriate crossover rule to the parent to produce a pair of offspring.

Step 4: Replace the parent with the offspring. Let $k = k + 1$.

Step 5: If $k > [p_C N_p]$, where p_C is a crossover rate, stop. Otherwise, go back to Step 2.

C. Mutation

Since the crossover produces offspring that only have characteristics from their parents, the manifold of the population is likely to be restricted within a narrow extent. A mutation operation can compensate this problem and keep the manifold by replacing the current allele with others with a given probability, say p_M. A simple flip–flop type mutation takes place such that: first select randomly an individual, select randomly a mutation point for the selected individual, reverse the bit thereat, and repeat until the number of this operation exceeds $[p_M N_p L]$. For example, when such a mutation point locates at the third place from the left-hand side, a change occurs for the gene of the selected individual.

$$\text{Before}:\quad 01(1)01101 \qquad \text{After}:\quad 01(0)01101$$

In addition to the above, varieties of mutation methods have been proposed so far. They are as follows:

1. Displacement: move part of the gene to another position of the same chromosome.
2. Duplication: copy some of the genes to another position.
3. Inversion: reverse the order of some genes in the chromosome.
4. Addition: insert some of the genes in the chromosome. This causes an increase in the length of the chromosome.
5. Deletion : delete some of the genes in the chromosome. This causes a decrease in the length of chromosome.

D. Summary of the Algorithm

The entire GA procedure is outlined in the following. The flow chart is shown in Figure 2.4.

Step 1: Let $t = 0$. Generate N_p individuals randomly and define the initial population $P_{OP}(0)$.

Step 2: Evaluate the fitness value for each individual. When $t = 0$, go to Step 3. Otherwise reproduce the individuals by applying an appropriate production rule.

Step 3: Under the prescribed probabilities, apply crossover and mutation in turn. These genetic operations produce the updated population $P_{OP}(t+1)$.

Step 4: Check the stopping condition. If it is satisfied, select the individual with highest fitness as a (near) optimal solution and stop. Otherwise, go back to Step 2 after letting $t := t + 1$.

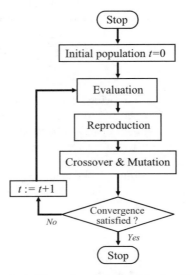

Fig. 2.4. Flow chart of the GA algorithm

Stopping conditions are commonly used as follows:

1. After a prescribed number of generations
2. When the highest fitness is not updated for a prescribed period
3. When the average fitness of the population has been almost saturated
4. A combination of the above conditions

Eventually, major factors of GA refer to the reproduction in Step 2 and the genetic operations in Step 3. In a word, the reproduction makes a point of finding better solutions and concentrates on the search around these, while the crossover and mutation try to spread the search space via a stochastic perturbation and to avoid staying at the local optimum. With a better complement of these properties with each other, GA can be used as an efficient search technique. Since GA is problem specific, it is necessary to adjust parameters such as mutation rate, crossover rate and population size to find reasonable settings for the problem class being worked on. A very small mutation rate may lead to genetic drift or premature convergence in a local optimum. On the contrary, a mutation rate that is too high may lead to the loss of good solutions.

E. Miscellaneous

The building block hypothesis is a theoretical background that supports the effectiveness of GA [13, 6]. It says that short, low order, and highly fit schemata are sampled, recombined, and re-sampled to form strings of potentially higher fitness. In a way, by working with these particular schemata (the building blocks), we have reduced the complexity of our problem; instead of building high-performance strings by trying every conceivable combination, we construct better and better strings from the best partial solutions of past samplings. This hypothesis requires coding to satisfy the following conditions.

- Individuals having similar phenotype are also close to each other regarding genotype.
- No major interference occurs between the loci.

From this aspect, Gray coding $(g_{l-1}, g_{l-2}, \ldots, g_0)$ is known to be more favorable than binary coding $(b_{l-1}, b_{l-2}, \ldots, b_0)$ because it can avoid the case where many simultaneous mutations or crossovers need to change the chromosome for a better solution. For example, let us assume that value 7 is optimal, and there exist the near optimal solutions with value 8. For these values, 4 bit binary cording of 7 is 0111 and 1000 for 8. Meanwhile, Gray coding becomes 0100 and 1100, respectively. Then, Gray coding can change 8 into 7 only by one mutation, but the binary coding needs such an operation four times successively. The following equation gives the relation between these types of coding:

$$g_k = \begin{cases} b_{l-1} & \text{if } k = l - 1 \\ b_{k+1} \oplus b_k & \text{if } k \leq l - 2 \end{cases},$$

where operator $\oplus$ applies the exclusive disjunction.

By virtue of the nature related to multi-start algorithms, we can expect to attain the global optimum more easily and more certainly than with any conventional single-start algorithms. To make use of this advantage, keeping the manifold during the search is a special importance for GA. In a sense, this is closely related to the status of the initial population and the stopping condition. The following are a few other well-known tips:.

1. The initial population should be selected by extracting the best N_p among the individuals with more than the prescribed population size ($> N_p$).
2. Mutation may destroy the favorable schema that crossover has built (building block hypothesis). Hence parameters controlling these operations should be set as $p_C > p_M$, and additionally p_M is designed so as to decrease along with the generation.

When applying GA to the constrained optimization problem described as

$$[Problem] \quad \min \ f(x) \ \text{subject to} \ \begin{cases} g_i(x) \geq 0 & (i = 1, 2, \ldots, m_1) \\ h_j(x) = 0 & (j = m_1 + 1, \ldots, m), \end{cases}$$

the following penalty function approach is usually adopted:

$$f'(x) = f(x) + P\{\sum_{i=1}^{m_1} \max[0, -g_i(x)] + \sum_{i=m_1+1}^{m} h_i(x)^2\},$$

where $P(> 0)$ denotes the penalty coefficient.

The real number coding is better and provides higher precision for the problem with a large search space where the binary coding would require a prohibitively long representation. This coding is straightforward, and the real value of each variable corresponds directly to the gene of each chromosome. The crossover is defined arithmetically as a linear combination of two vectors. When P_1 and P_2 denote the parent solution vectors, the offspring are generated as $O_1 = aP_1 + (1-a)P_2$ and $O_2 = (1-a)P_1 + aP_2$, where a is a random value in $\{0, 1\}$. On the other hand, the mutation starts with randomly selecting an individual V. Then the mutation is applied in two ways, that is, simple mutation applies the following equation only for mutation point k appointed randomly in V, while the uniform mutation applies this to every locus:

$$V_k = v_k^L + r(v_k^U - v_k^L) \ \text{where} \ k = \begin{cases} \exists k : & \text{for simple mutation} \\ \forall k : & \text{for uniform mutation} \end{cases},$$

where v^U and v^L are the lower and upper bounds, respectively, and r is a random number from uniform probability distribution. A certain local search scheme is generally incorporated for these genetic operations to find a better solution near the current one.

2.2.2 Simulated Annealing

Simulated annealing (SA) is another metaheuristic algorithm specially suitable for the global optimization in terms of giving a certain probabilistic perturbation [7, 8]. It borrows the idea from a physical mechanism known as annealing in metallurgy. Annealing is a popular engineering technique that applies heating and controlled cooling for material to increase the size of its crystals and reduce their structural defects. Heating causes the atoms to activate the kinetic energy, and is likely to make them unstuck at their initial positions (a local minimum of the internal energy) and wander randomly through states of higher energy. In contrast, slow cooling gives atoms more chances of finding configurations with lower internal energy than the initial one.

By analogy with this physical process, SA tries to solve the optimization problem. In its solution, each point of the search space is compared to a state of some physical system, and the objective function to be minimized is interpreted as the internal energy status of the system. When the system attains the state with the minimum energy, we can claim that the optimal solution has been obtained. Its basic iteration process is described as follows.

Step 1: Generate an initial solution (let it be a current solution x), and also set an initial temperature T.

Step 2: Consider some neighbors of the current state, and select randomly a neighbor x' in it as a possible solution.

Step 3: Decide probabilistically whether to move on state x' or to stay at state x.

Step 4: Check the stopping condition, and if it is satisfied, stop. Otherwise, cool the temperature and go back to Step 2.

The probability moving from the current solution to the neighbor in Step 3 depends on the difference between the respective objective function values and a time-varying parameter called temperature T. The algorithm is designed so that the current solution changes almost randomly when T is high, while the solution descends downhill as a whole with the decrease in temperature. The allowance for uphill moves during the process may avoid sticking at the local minima and make it possible to be a good approximation of the global optimum as illustrated in Figure 2.2. Let us describe the detail of the essential features of SA in the following.

A. Neighbors of State

Though the selection of neighbors (local search) has a great affect on the performance of the algorithm, no general methods have been proposed since they are very problem-specific. The concept of local search may be modeled conveniently as a search graph where vertices represent the states, and an edge denotes a transition between the vertices. Then, the length of a path represents the degree of the niche of neighbors, supposing the neighbors are expected to all have nearly the same energy. It is desirable to go from the initial state to a better state successively by a relatively short path on this graph, and such a path must be followed by the iteration of SA as similarly as possible.

Regarding the generation of neighbors, many ideas have been proposed for each class of problem so far, *i.e.*, the n-opt neighborhood and the *or*-opt neighborhood in the traveling salesman problem; the insertion neighborhood and the swap neighborhood in the scheduling problem; the λ-flip neighborhood in the maximum satisfiability problem, and so on [1].

B. Transition Probabilities

The transition from the current state x to a candidate state x' will be made according to the probability given by a function $p(e, e', T)$ where $e = E(x)$ and $e' = E(x')$ denote the energies of the two states (presently objective function values). An essential requirement for the transition probability is that $p(e, e', T)$ is non-zero when $e' \geq e$. This means that the system may move to the new state even if it is worse (has a higher energy) than the current one. The allowance for such uphill moves during the process may avoid sticking at the local minimum, and one can expect a good approximation of the global optimum as noted already.

On the other hand, as T tends to zero, the probability $p(e, e', T)$ also approaches zero when $e' \geq e$, while keeping a reasonable positive value when $e' < e$. As T becomes smaller and smaller; therefore, the system will increasingly favor downhill moves, and avoid the uphill moves. When T approaches 0, SA performs just like the greedy algorithm, which makes the move if and only if it goes downhill.

The probability function is usually chosen so that the probability of accepting a move decreases according to the increase in the difference of energies $\Delta e = e' - e$. Moreover, the evolution of x should be sensitive to Δe over a wide range when T is high and only within the small range when T is small. This means that small uphill moves are more likely to occur than large ones in the latter part of the search. To meet such requirements, the following Maxwell–Boltzmann distribution governing the distribution of energies of molecules in a gas is popularly used (see also Figure 2.5):

$$p = \begin{cases} 1 & \text{if } \Delta e \leq 0 \\ \exp(-\Delta e / T) & \text{if } \Delta e > 0 \end{cases}.$$

C. Annealing Schedule

Another essential feature of SA is how to reduce the temperature gradually as the search proceeds. This procedure is known as the annealing (cooling) schedule. Simply speaking, the initial temperature is set to a high value so that the uphill and downhill transition probabilities become nearly the same. To do this, it is necessary to estimate Δe for a random state and its neighbors over the entire search space. However, this needs some amount of preliminary experiments. A more common method is to decide the initial temperature so

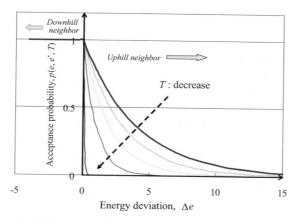

Fig. 2.5. The demanding character of probability function

that the acceptance rate in the search at the earlier stage will be greater than a prescribed value.

The temperature must decrease to zero or nearly zero by the end of the iteration. This is the only condition required for the cooling schedule, and many methods have been proposed so far. Among them, geometric cooling is a simple but popular method, in which the temperature is decreased by a fixed rate at each step, *i.e.*, $T := \beta T$, $(\beta < 1)$. Another one termed exponential cooling is applied as $T = T_0 \exp(-at)$, where T_0 and t denote an initial temperature and iteration number, respectively. A more sophisticated method involves a heat-up step when the tentative solution has not been updated at all during a certain period. By returning the current temperature to the previous one, it tries to break the plateau status. In this way, the initial search makes a point of wandering a broad space that may contain good solutions while ignoring small degradations of the objective function. Then it will drift towards the low energy regions that become narrower and narrower, and finally aim at the minimum according to the descent strategy.

D. Convergence Features

It is known that the probability of finding the global optimal solution by SA approaches 1 as the annealing schedule is continued infinitely. This theoretical result is not helpful for deciding a stopping condition in practice. The simplest condition is to terminate the iterations after the prescribed number for which the temperature is reduced nearly by zero according to the annealing schedule. Various methods can be considered by observing the status of convergence more elaborately in terms of an update of the tentative solution.

Sometimes it is better to move back to a solution that was significant rather than always moving from the current state. This procedure is called restarting. The decision to restart could be made based on a fixed number of steps, or on the current solution being too poor compared with the best one obtained so far.

Finally, applying SA to a specific problem, we must specify the state space, the neighbor selection method, the probability transition function, and the annealing schedule. These choices can have a significant impact on the effectiveness of the method. Unfortunately, however, there is neither specific value that will work well with all problems, nor a general way to find the best setting for a given problem.

2.2.3 Tabu Search

Though tabu search (TS) [9, 10] has a simple solution procedure, it is an effective method for combinatorial optimization problems. In a word, TS belongs to a class of local search techniques that enhances performance by using a special memory structure known as the tabu list. TS repeats the local search iteratively to move from a current solution x to a possible and best solution x' in the neighbor of $x, N(x)$. Unfortunately, there exists the case where simple local search may cause a cycling of the solution, $i.e.$, from x to x', and from x' to x. To avoid such cycling, TS use the tabu list that corresponds to a short term memory cited in the field of recognition science. Transition to any solutions involved in the tabu list is prohibited for a while, even if this will provide an improvement of the current solution. Under such restrictions, TS continues the local search until a certain stopping condition has been satisfied. The basic iteration process is outlined as follows:

Step 1: Generate an initial solution x and let $x^* := x$, where x^* denotes the current best solution. Set $k = 0$ and let the tabu list $T(k)$ be empty.

Step 2: If $N(x) - T(k)$ is empty, stop. Otherwise, set $k := k + 1$ and select x' such that $x' = \min f(x)$ for $\forall x \in N(x) - T(k)$.

Step 3: If x' outperforms the current solution x^*, $i.e.$, $f(x') \leq f(x^*)$, let $x^* := x'$.

Step 4: If a chosen number of iterations has elapsed either in total or since x^* was last improved, stop. Otherwise, update $T(k)$, and go back to Step 2.

In the TS algorithm, the tabu list plays the most important role. It makes it possible to explore the search space that would be left unexplored and to escape from the local optimum. The simplest form of the tabu list is a list of the solutions by the latest m-visits. Referring to this list, transition to the solutions recorded in the tabu list is prohibited to move during a period of m length. Such period is called the tabu tenure. In other words, the validity of such prohibition holds only during the tabu tenure, and its length can control the regulation regarding the transition. That is, if it is long, then the transition is hardly restricted and *vice versa*.

Other structures of the tabu list utilize certain attributes associated with the particular search technique depending on the problem. Solutions with such attributes are labeled to be tabu-active, and the tabu-active solutions are also viewed as tabu for the search. For example, in the traveling salesman problem (TSP), solutions that include certain arcs are prohibited or an arc that was added newly to a TSP tour cannot be removed in the next m-moves. Generally speaking, tabu lists containing the attributes are much more effective. However, by forbidding the solutions that contain tabu-active elements, more than one solution is likely to be declared as the tabu. Hence, there exist the cases where some solutions might be avoided although they have excellent quality and have not yet been visited.

Aspiration criteria serve to relax such restrictions. They allow overriding the tabu state of the solutions that are better than the currently best known solution, and keep it in the allowed set. Besides these special ideas, a variety of extensions are known, some of which are cited below.

The load of local search can be reduced if we concentrate the search only on the promising extent instead of whole neighbor. Such an idea is generally called a candidate list strategy. After selecting k-best solutions among the neighbor solutions, probabilistic tabu search is to replace the current solution randomly with one depending on the probabilities, which are decided based on their objective functions. This idea is very similar to the roulette strategy in the selection of GA.

In addition to the function of the tabu list as a short-term memory, a long-term memory is available to improve the performance of the algorithm. This generic name refers to an idea that tries to utilize the history of information along with the search process. The long-term memory makes it possible to use the intensification of promising search and diversification for global search at the same time. For example, a transition measure in frequency-based memory records the numbers of the modification of the variables, while a residence measure records the number staying at the specific value. Since the high transition measure foresees the long-term search cycle, an appropriate penalty should imposed on its selection. On the other hand, the residence measure is available for the selection of initial solutions by controlling the appearance rate of the certain variables. That is, restriction of the variables with high measure can facilitate the diversification while promoting the intensification.

2.2.4 Differential Evolution (DE)

Differential Evolution (DE) is viewed as a real number coding version of GA and was developed by Price and Storn [11]. Though it is a very simple population-based optimization method, it is known as a very powerful method for real world applications. A variety of variants are classified using a triplet expression like DE/x/y/z/, where

- x specifies the method for selecting the parent vector to become a base of the mutant vector. Two selections, *i.e.*, chosen randomly ("rand") or

chosen from the best in the current population ("best") are typically employed.
- y is a number of the difference vector used in Equation 2.2.
- z denotes a crossover method. In binominal crossover ("bin"), crossover is performed on each gene of a chromosome while, in exponential crossover ("exp") it is performed on a chromosome as a whole:

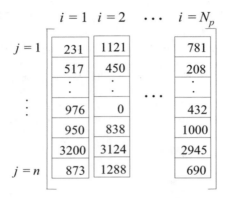

$$i = 1 \quad i = 2 \quad \cdots \quad i = N_p$$

	$i=1$	$i=2$		$i=N_p$
$j=1$	231	1121		781
	517	450		208
	.	.		.
$\vdots$	976	0		432
	950	838		1000
	3200	3124		2945
$j=n$	873	1288		690

where N_p = population size
n = number of decision variables

Fig. 2.6. Example of coding in DE

As is usual with every variant, users need the following settings before optimizing their own problem: the number of population N_p, scaling factor F and crossover rate p_C. The algorithm in the case of DE/rand/1/bin/ is outlined as follows:

Step 1 (Generation): Generate randomly every n-dimensional "target" vector to yield the initial population.

$$P_{\mathrm{OP}}(t) = \{x_{i,t}\} \ (i = 1, 2, \dots, N_p),$$

where t is a generation number and N_p is a population size. An example of coding is shown in Figure 2.6.

Step 2 (Mutation): Create each "mutant" vector by adding the weighted difference between two target vectors to the third target vector. These three vectors are chosen randomly among the population,

$$v_{i,t+1} = x_{r3,t} + F(x_{r2,t} - x_{r1,t}) \ (i = 1, 2, \dots, N_p), \tag{2.2}$$

where F is real and constant in $[0, 2]$.

Step 3 (Crossover): Apply the crossover operation to generate the trial vector u_i by mixing some elements of the target vector with the mutant vector through comparison between the random value and the crossover rate (see also Figure 2.7),

$$u_{ji,t+1} = \begin{cases} v_{ji,t+1} & \text{if } \text{rand}(j) \le p_C \text{ or } j = \text{rand}() \\ x_{ji,t} & \text{if } \text{rand}(j) > p_C \text{ and } j \ne \text{rand}() \end{cases} \quad (j = 1, 2, \ldots, n),$$

where $\text{rand}(j)$ is the j-th evaluation of a uniform random number generator, p_C is the crossover rate in $[0, 1]$, and $\text{rand}()$ is a randomly chosen index in $\{1, 2, \ldots, n\}$. Ensure that $u_{i,t+1}$ has at least one elements from the mutant vector $v_{i,t+1}$. Then evaluate the performance of each vector.

Step 4 (Selection): If the trial vector outperforms the target vector, the target vector is replaced with the trial vector. Otherwise, the target vector is retained. Thus, the members of the new population for the next generation are selected in this step.

Step 5: Check the stopping condition. If it is satisfied, stop and return the overall best vector as the final solution. Otherwise, go back to Step 2 by incrementing the generation number by 1.

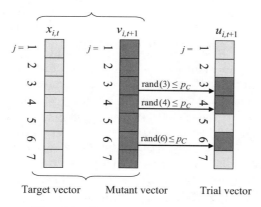

Fig. 2.7. Crossover operation of DE

In the case of DE/best/2/bin, at the above Step 2, the mutant vector is derived from the following equation:

$$v_{i,t+1} = x_{\text{best},t} + F(x_{r1,t} + x_{r2,t} - x_{r3,t} - x_{r4,t}) \quad (i = 1, 2, \ldots, N_p),$$

where $x_{\text{best},t}$ is the best solution at generation t. Moreover, the exponential crossover in Step 3 is applied as

$$u_{ji,t+1} = \begin{cases} v_{ji,t+1} & \text{if rand}() \leq p_C \\ x_{ji,t} & \text{if rand}() > p_C \end{cases} \quad \text{(for } \forall j\text{)}.$$

For successful application of DE, there are several tips regarding parameter setting and tuning, some of which will be shown below.

1. The number of population N_p is normally set between five to ten times the number of decision variables.
2. If a proper convergence cannot be attained, it is better to increase N_p, or adjust F and p_C both in the range $[0.5, 1]$ for most problems.
3. Simultaneous increase in N_p and decrease in F make the convergence more likely to occur but generally make it longer.
4. DE is much more sensitive to the choice of F than p_C. Though larger p_C gives faster convergence, it is sometimes necessary to use a smaller value to make DE robust enough for the particular problem. Thus, there is always a tradeoff between convergence speed and robustness.
5. p_C of binominal crossover should usually be set higher than that of the exponential crossover.

A. Adaptive DE

To improve the convergence, a variant of DE (ADE) was proposed recently[1] . It introduced ideas of a gradient field in the objective function space and an age for individuals to control the crossover factor. The algorithm is outlined below.

Step 1(Generation): Reset the generation at 1 and the age at 0. $Age(i)$ is defined as the number of generations during which each individual i is alive. Then generate $2N_p$ individuals x_i in n-dimensional space.

Step 2 (Gradient field): Make a pair randomly for each individual and compare their objective function values. Then, classify them into winner (having smaller value) and loser, and register as winner and loser, respectively. The winners will age by one, and the losers rejuvenate by one.

Step 3 (Mutation): Pick up randomly a base vector $x_{\text{base}()}$ from the winner. Moreover, choose randomly a pair building the gradient field and generate a mutant vector as follows:

$$v_{i,t+1} = x_{\text{base},t} + F(x_{\text{better}(),t} - x_{\text{worse}(),t}) \quad (i = 1, \ldots, 2N_p),$$

where $x_{\text{better}()}$ and $x_{\text{worse}()}$ denote the winner and loser of each pair, respectively. This operation may generate mutants in the direction possible for decreasing the objective function globally everywhere in the search space.

---

[1] Shimizu Y (2005) About adaptive DE. *Private Communication*

Step 4 (Crossover): The same type of crossover as has already been mentioned is available. However, its rate p_C will be decided by a monotonic decreasing function of age, e.g.,

$$p_C = (a + c)e^{-b \cdot Age(i)} + c, \quad \text{or} \quad p_C = \max[a + c - b \cdot Age(i), \ c],$$

where a, b and c are real positive constants to be determined by the user under the condition that $0 < a+c < 1$ (see 2.8). This crossover rate makes the target vectors that have lived for long time (having an older age) more likely to survive in the next generation.

Step 5 (Selection): If the trial vector is better than the target vector, replace the target vector with the trial vector and give it a new age suitably e.g., reset (0). Otherwise, the target vector is retained and it gets older by one.

Step 6: Check the stopping condition. If it is satisfied, stop and return to the overall best vector as the final solution. Otherwise, go back to Step 2 by updating the generation.

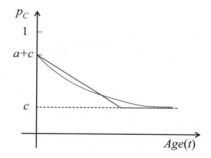

Fig. 2.8. Crossover rate depending on age

The following simple test problem validates the effectiveness of this method. Minimization of the Rosenbrock function is compared with the conventional method DE/rand/1/bin/:

$$f(x) = 100 \cdot (x_1^2 - x_2)^2 + (1 - x_1)^2, \quad x_1, \ x_2 \in [-10, 10].$$

Although, there are only two decision variables, this problem has the reputation of being a difficult minimization problem. The global minimum is located at $(x_1, x_2) = (1, 1)$. The comparison of convergence features between ordinal and adaptive DE is shown in Figure 2.9 in the logarithm scales. The linear model of age is used to calculate p_C as $p_C = 0.5 \cdot \max[1 - 0.0001 \cdot Age(i), 0.5]$. The adaptive method ("DE-rev") is known to present a good convergence feature compared with the conventional method ("DE-org").

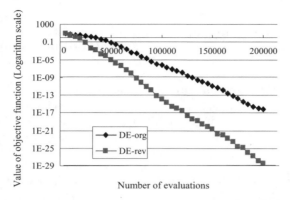

Fig. 2.9. Comparison of convergence features

2.2.5 Particle Swarm Optimization (PSO)

Particle Swarm Optimization (PSO) , which was developed by J. Kennedy [12], is also a real number coding metaheuristic method for optimization. It is a form of swarm intelligence in the artificial intelligence study of the collective behavior in decentralized and self-organized systems. It stems from the theory of boids by C. Reynolds [13]. Imagining the behavior of a swarm of insects or a school of fish, we can observe that when one member finds a desirable path to go, (*i.e.*, for food, protection, *etc.*), the rest of the swarm can follow it quickly even if they are on the opposite side of the swarm.

The algorithm of PSO relies on the strength that such behavior to attain the goal is rational, and can be simulated by only three movements termed separation, alignment, and cohesion.

- Separation is a rule to separate one object from a neighbor, and prevent from colliding with each other. For this purpose, a boid flying ahead must speed up while those in the rear slow down. Moreover, the boids can change direction to avoid obstacles.
- By alignment, all objects try to adapt their movement to the others. Front boids flying far away will slow down and the rear boids will speed up to catch up.
- Cohesion is a centripetal rule for not disturbing the shape of the population as a whole. This requires boids to fly to the center of the swarm or the gravity point.

According to these three movements, PSO can be developed by imaging boids with a position and a velocity. These boids fly through hyperspace and remember the best position that they have seen. Members of a swarm communicate with each other and adjust their own position and velocity based on the information regarding the good positions both of their own (local bests)

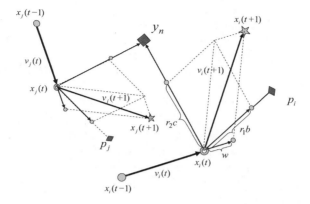

Fig. 2.10. Search scheme of PSO

and a swarm best (global best) as depicted in Figure 2.10. Updating of the position and the velocity is done through the following formulas:

$$x_i(t+1) = x_i(t) + v_i(t+1), \tag{2.3}$$
$$v_i(t+1) = w \cdot v_i(t) + r_1 b(p_i - x_i(t)) + r_2 c(y_n - x_i(t))$$
$$(i = 1, 2, \ldots, N_p), \tag{2.4}$$

where

t is the generation,
N_p is the population size (number of boids),
w is an inertial constant (usually slightly less than 1),
b and c are constants making a point of how much the boid is directed toward
    the good position (usually around 1),
r_1 and r_2 are random values in the range [0,1],
p_i is the best position seen by the boid i,
y_n is the global best position seen by the swarm.

The algorithm is outlined below.

Step 1: Set $t = 1$.
    Initialize $x(t)$ and $v(t)$ randomly within the range of these values.
    Initialize each p_i to the current position.
    Initialize y_n to the position that has the best fitness among swarms.
Step 2: For each boid, do the following:
    obtain $v_i(t+1)$ according to the Equation 2.4,
    obtain $x_i(t+1)$ according to the Equation 2.3,

evaluate the new position,
if it outperforms p_i, update it,
if it outperforms y_n, update it.

Step 3: If the stopping condition is satisfied, stop. Otherwise let $t := t + 1$, and go back to Step 2.

2.2.6 Other Methods

In what follows, a few useful methods will be introduced. Generally speaking, they can exhibit advantages over the methods mentioned above for a particular class of problems. Moreover, they are amenable for various hybrid approaches of metaheuristic methods relying on the features characterized by probabilistic deviation, multi-modality, population-base, multi-start, etc.

The ant colony algorithm (ACO) [14, 15] is a probabilistic optimization technique that mimics the behavior of ants finding paths from the colony to food. In nature, ants wander randomly to find food. On the way back to their colony, they lay down pheromone trails. If other ants find such trails, they can reach the food source more easily by following the trail. Hence, if one ant can find a good or short path from the colony to the food source, other ants are more likely to follow that path. Since the pheromone trail evaporates with time, its attractive strength will gradually reduce. The more time it takes for an ant to travel, the more pheromones will evaporate. Since a short path is traced faster, the pheromone density remains high. Such positive feedback eventually makes all the ants follow a single path. Pheromone evaporation has also the advantage of avoiding the convergence to a local optimum. ACO has an advantage over SA and GA when the food source may change dynamically, since it can adapt to the changes continuously. Moreover, this idea is readily available for applying a multi-start technique in various metaheuristic optimizations.

Memetic algorithm [16] is an approach emerging from traditional GA. By combining local search with the crossover operator, it can provide considerably faster convergence, say orders of magnitude, than traditional GA. For this reason, it is called genetic local search or the hybrid genetic algorithm. Moreover, it should be noticed that this algorithm is most suitable for parallel computing.

An evolutionary approach called scatter search [17] is very different from the other evolutionary methods. It possesses a strategic design mechanism to generate new solutions while other approaches resort to randomization. For example, in GA, two solutions are randomly chosen from the population and crossover or a combination mechanism is applied to generate one or more offspring. Scatter search works based on a set of solutions called the reference set, and combines these solutions to create new ones based on the generalized path constructions in Euclidean space. That is, by both convex (linear) and non-convex combination of two different solutions, the reference set can evolve in turn (reference set update)[2].

In Figure 2.11 it is assumed that the original reference solution set consists of the circles labeled A, B and C (diversified generation, enhancement). In terms of a convex combination of reference solutions A and B (solution combination), a number of solutions in the line segment defined by A and B may be created (subset generation). Among them, only solution 1 that satisfies a certain criteria for membership is involved in the reference set. In the same way, convex and non-convex combinations of original and new reference solutions create points 2, 3 and 4, one after another. After all, the resulting reference set consists of seven solutions in the present case. Unlike a "population" in GA, the number of reference solutions is relatively small in scatter search. Scatter search chooses only two or more reference solutions in a systematic way to create new solutions as shown above.

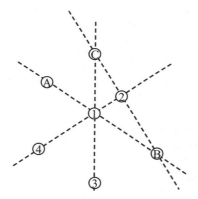

Fig. 2.11. Successive generation of solutions by scatter search

The following five major features characterize the implementation of scatter search.

1. Diversified generation: to generate a set of diverse trial solutions using an arbitrary initial solution (or seed solution).
2. Enhancement: to transform a trial solution into one or more improved trial solutions.
3. Reference set update: to build and maintain a reference set consisting of the k-best solutions found (where the value of k is typically small, $e.g.$, no more than 20). Solutions gain membership to the reference set according to their quality or their diversity.
4. Subset generation: to produce a subset of its solutions as a basis for creating combined solutions.

---

[2] This is similar to the movement of the simplex method stated in Appendix B.

5. Solution combination: to transform a given subset of solutions into one or more combined solution vectors.

In the sense that this method will rely on the reference solutions, this idea can also be used for applying the multi-start technique in some metaheuristic approaches.

2.3 Hybrid Approaches to Optimization

Since the term "hybrid" has broad and manifold meanings, we can give several hybrid approaches even if discussion might be restricted within the optimization methods. In what follows, three types of hybrid approach will be presented in terms of the combination of traditional mathematical programming (MP) and recent metaheuristic optimization (meta).

The first category is a "MP–MP" class. Most gradient methods for multi-dimensional optimization involve the optimization of step size search along the selected direction in the course of iteration. For this search, a scalar optimization method like the golden section algorithm or the Fibonatti algorithm is commonly used. This is a plain example of the hybrid approach in this class. Using an LP-relaxed solution as an initial solution and applying nonlinear programs (NLP) at the next stage may be another example of this class.

The second class "meta–meta" mainly appears in the extended or sophisticated application of the original algorithm of the metaheuristic method. Using the ACO method as the restarting technique of another metaheuristic method is an example of this class. Combining a binary code GA with other real number coding meta-methods is a reasonable way to cope with mixed-integer programs (MIP) . Instead of applying each method individually to solve MIP, such a hybrid approach can bring about a synergic effect to reduce the search space (chromosome length) and to improve the accuracy of the resulting solution (size of grains or quantification).

After all, many practical hybrid approaches may belong to the third "meta–MP" class. As supposed from the memetic algorithm or genetic local search, the local search is considered to be a promising technique that can accelerate the efficiency of the search compared with the single use of the metaheuristic method. Every method using an appropriate optimization technique for such local search may be viewed as a hybrid method in this class.

A particular advantage of this class will be exhibited to solve the following MIP in a hierarchical manner:

$[Problem] \quad \min_{x,z} f(x,z)$

$$\text{subject to} \left\{ \begin{array}{ll} g_i(x,z) \geq 0 & (i=1,2,\ldots,m_1) \\ h_i(x,z) = 0 & (i=m_1+1,\ldots,m) \\ x \geq 0, & \text{(real)} \\ z \geq 0, & \text{(integer)} \end{array} \right. .$$

This approach can achieve a good match not only between the upper and lower level problems but also each problem and the respective solution method. The most serious difficulties in solving MIP problems refer to the combinatorial nature in solution. By pegging the integer variables at the values decided at the upper level, the resulting lower level problem is reduced to a usual (non-combinatorial) problem that it is possible to be solved reasonably by MP. On the other hand, the upper level problem becomes an unconstrained integer programs (IP), and it is treated effectively by the metaheuristic method. Based on such an idea, the following hierarchical formulation is known to be amenable to solving MIP in a hybrid manner of "meta-MP" type (see also to Figure 2.12):

$$[Problem] \quad \min_{z \geq 0:\text{integer}} \quad f(x,z)$$

$$\text{subject to} \quad \min_{x \geq 0:\text{ real}} \quad f(x,z),$$

$$\text{subject to} \left\{ \begin{array}{ll} g_i(x,z) \geq 0 & (i=1,\ldots,m_1) \\ h_i(x,z) = 0 & (i=m_1+1,\ldots,m) \end{array} \right. .$$

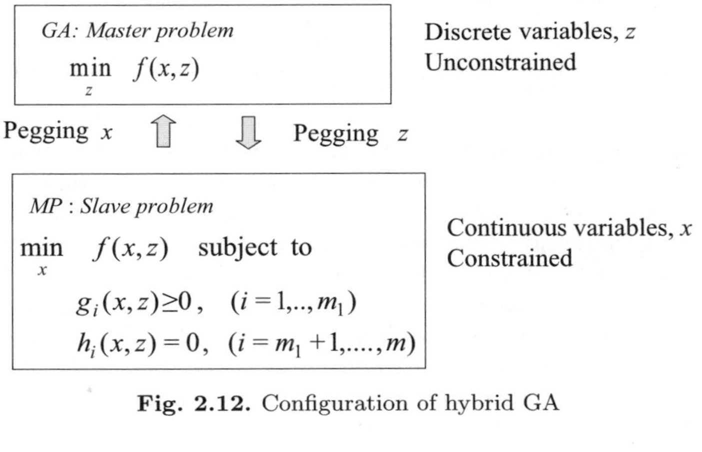

Fig. 2.12. Configuration of hybrid GA

In the above, the lower level problem becomes the usual mathematical programming problem. When the constraints of pure integer variables are involved, a penalty function method is available at the upper level as follows:

$$\min_{z \geq 0:\text{integer}} \quad f(x,z) + P\{\sum_i\} \max[0, -g_i(z)] + \sum_i h_i(z)^2\}.$$

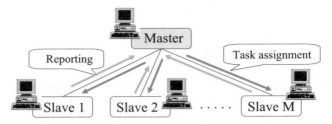

Fig. 2.13. Master–slave configuration for parallel computing

Moreover, by noticing the analogy of the above formulation to the parallel computing of the master–slave configuration as shown in Figure 2.13, an effective parallel computing is readily implemented [32]. There are many combinatorial optimization problems formulated as IP and MIP at every stage of the manufacturing optimization. The scheme presented here has close connections to various manufacturing optimization problems for which we can deploy this approach in an effective manner. For example, a large-scale network design and a site location problem under multi-objective optimization will be developed in the following sections.

2.4 Applications for Manufacturing Planning and Operation

Recent innovations in information technology as well as advanced transportation technologies are accelerating globalization of markets outstandingly. This raises the importance of just-in-time and agile manufacturing much more than before, since its effectiveness is pivotal to the efficiency of the business process. From this point of view, we will present three applications ranging from strategic planning to operational scheduling. We will also show how effectively the optimization problem in each topic can be solved by the relevant method employed there.

The first topic takes a logistic problem associated with supply chain management (SCM) [19, 20, 21]. It will be formulated as a hub facility location and route selection problem attempting to minimize the total management cost over the area of interest. This kind of problem [22, 23, 24] is also closely related to the network design of hub systems popular in various fields such as transportation [25], telecommunication [26], *etc.* However, most previous studies have scarcely called attention to the entire system composed both of distribution and collection networks. To deal with such large-scale and complex problems practically, an approach that decomposes the problem into sub-problems and applies a hybrid tabu search method will be described [27].

In terms of the small-lot-multi-kinds production, the introduction of mixed-model assembly lines is becoming popular in manufacturing. To increase the efficiency of such line handling, it is essential to prevent various line stoppages incurred due to unexpected inconsistencies [28, 29]. The second topic concerns an injection sequencing problem for the manufacturing represented by the car industry [30]. The mixed-model assembly line thereat includes a painting line where we need to pay attention to uncertainties associated with so-called defective products. After formulating the problem, SA is employed to solve the resulting combinational optimization problem in a numerically effective manner.

The scheduling problem is one of the most important problems associated with the effective operation of manufacturing systems. Consequently, much of research has been done [31, 32, 33, 34], but most work only describes simple models [35]. Additionally, it should be noticed that the roles of human operators are still important although automation is now becoming popular in manufacturing. However, little research has taken into account the role of operators and the cooperation between operators and resources [36]. The third topic concerns a production scheduling managed by multi-skilled human operators who can manipulate multiple types of resources such as machine tools, robots, and so on [37]. After formulating a general scheduling problem associated with human tasks, a practical method based on a dispatching rule or an empirical optimization will be presented.

2.4.1 Logistic Optimization Using Hybrid Tabu Search

Recently, industries have been paying keen attention to SCM and studying it from various aspects [38, 39, 40]. It is viewed as a reengineering method managing life cycle activities of a business process to deliver added-value products and service to customers. As an essential part of decision making in such business processes, we consider a logistic optimization associated with a supply chain network(SCN) [27]. It is composed of suppliers, collection centers (CCs), plants, distribution centers (DCs), and customers as shown in Figure 2.14. Though CC can receive materials from multiple suppliers due to risk aversion (multiple allocation), each customer will receive products only from one DC (single allocation) that can deliver products either from another DC or customer. The problem is formulated under the conditions that the capacity of the facility is constrained, and demand, supply and per unit transport cost are given *apriori*. It refers to a nonlinear mixed-integer programming problem (MINLP) simultaneously deciding the location of hub centers and routes to meet the demands of all SCN members while minimizing the total cost,

$$\min \quad \sum_{i \in I} \sum_{j \in J} D_i C1_{ij} r_{ij} + \sum_{j \in J} \sum_{j' \in J} \left(\sum_{i \in I} D_i r_{ij} \right) C2_{jj'} s_{jj'}$$

$$+ \sum_{j' \in J} \sum_{k \in K} \left(\sum_{j \in J} \left(\sum_{i \in I} D_i r_{ij} \right) s_{jj'} \right) C3_{j'k} t_{j'k}$$

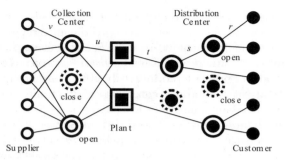

Fig. 2.14. Supply chain network

$$+ \sum_{k \in K} \sum_{l \in L} C4_{kl} u_{kl} + \sum_{l \in L} \sum_{m \in M} C5_{lm} v_{lm} + \sum_{j \in J} F1_j x_j + \sum_{l \in L} F2_l y_l,$$

subject to

$$\sum_{j \in J} r_{ij} = 1, \quad \forall i \in I, \tag{2.5}$$

$$\sum_{i \in I} D_i r_{ij} \leq P_j x_j, \quad \forall j \in J, \tag{2.6}$$

$$\sum_{j' \in J} s_{jj'} = x_j, \quad \forall j \in J, \tag{2.7}$$

$$\sum_{j \in J} \left(\sum_{i \in I} D_i r_{ij} \right) s_{jj'} \leq P_{j'} s_{j'j} x_{j'}, \quad \forall j' \in J, \tag{2.8}$$

$$\sum_{k \in K} t_{j'k} = s_{j'j'}, \quad \forall j' \in J, \tag{2.9}$$

$$\sum_{j' \in J} \left(\sum_{j \in J} \left(\sum_{i \in I} D_i r_{ij} \right) s_{jj'} \right) t_{j'k} \leq Q_k, \quad \forall k \in K, \tag{2.10}$$

$$\sum_{l \in L} u_{kl} = \sum_{j' \in J} \left(\sum_{j \in J} \left(\sum_{i \in I} D_i r_{ij} \right) s_{jj'} \right) t_{j'k}, \quad \forall k \in K, \tag{2.11}$$

$$\sum_{k \in K} u_{kl} \leq s_l y_l, \quad \forall l \in L, \tag{2.12}$$

$$\sum_{k \in K} u_{kl} = \sum_{m \in M} v_{lm}, \quad \forall l \in L, \tag{2.13}$$

$$\sum_{l \in L} v_{lm} \leq T_m, \quad \forall m \in M, \tag{2.14}$$

$r, \ s, \ t \in \{0, 1\}, \quad x, \ y \in \{0, 1\}, \quad u, \ v \in$ real number, where binary variables x_i and y_i take 1 if each center i is open, and r_{ij}, s_{ij}, t_{ij} become 1 if there exist routes between customer i and DC j, DC i and DC j, and DC i and plant j, respectively. Otherwise, they are equal to 0 in all cases. u_{ij} and v_{ij} denote the amount of shipping from CC j to plant i and from supplier j to CC i, respectively. Moreover, D_i is the demand of customer i, and P_i, Q_i, S_i and T_i represent capacities of DC, plant, CC and supplier, respectively.

On the other hand, the first to fifth terms of the objective function are related to transport costs while the sixth and seventh terms to fixed charge costs of DC and CC, respectively. Equations 2.5, 2.7, and 2.9 mean that each customer, DC and plant are allowed to select only one location each in the downstream network. Equations 2.6, 2.8, 2.10, 2.12, and 2.14 represent the capacity constraints on the first stage DCs and the second stage DCs, plant, CC, and supplier, respectively. Equations 2.11 and 2.13 represent balance equations between input and output of plant and CC, respectively.

A. Hierarchical Procedure for Solution

(1) Decomposition into Sub-models

Since the solution of MINLP belongs to an NP-hard class, developing a practical solution method is more desirable than aiming at a rigid optimum. Noting the particular structure of the problem as illustrated in Figure 2.15, we can decompose the original SCN into two sub-networks originating from the plants in opposite direction to each other, *i.e.*, upstream (procurement) chain, and downstream (distribution) chain. The former solves a problem of how to supply raw materials from suppliers to plants via CCs, while the latter concerns how to distribute the products from plants to customers via DCs.

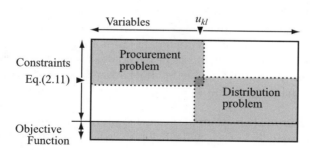

Fig. 2.15. A pseudo-block diagonal pattern of the problem structure

Eventually, to obtain a consequent result for the entire supply chain from what is solved individually, it is necessary to combine them consistently by adjusting a coupling constraint effectively. Instead of using Equation 2.11

directly as the coupling constraint, it is transformed into a suitable condition so that the tradeoff between the sub-networks can be adjusted through an auction-like mechanism based on an imaginary cost. For this purpose, we define the optimal cost associated with the procurement in the upstream chain C^*_{proc},

$$\sum_{k \in K} \sum_{l \in L} C4_{kl} u_{kl} + \sum_{l \in L} \sum_{m \in M} C5_{lm} v_{lm} + \sum_{l \in L} F2_l y_l = C^*_{\text{proc}}. \qquad (2.15)$$

Then, dividing C^*_{proc} into each plant according to the amount of production, $i.e.$, $C^*_{\text{proc}} = \sum_k V_k$, we view V_k as an estimated shipping cost from each plant. Then, by denoting the unit procurement cost by ρ_k, we obtain the following equation:

$$\rho_k \sum_{j' \in J} \left(\sum_{j \in J} \left(\sum_{i \in I} D_i r_{ij} \right) s_{jj'} \right) t_{j'k} = V_k, \quad \forall k \in K. \qquad (2.16)$$

Using this as a coupling condition instead of Equation 2.11, we can decompose the entire model into each sub-model as follows.
Downstream network (DC) model[3]

$$\min \quad \sum_{i \in I} \sum_{j \in J} D_i C1_{ij} r_{ij} + \sum_{j \in J} \sum_{j' \in J} \left(\sum_{i \in I} D_i r_{ij} \right) C2_{jj'} s_{jj'}$$

$$+ \sum_{j' \in J} \sum_{k \in K} \left(\sum_{j \in J} \left(\sum_{i \in I} D_i r_{ij} \right) s_{jj'} \right) C3_{j'k} t_{j'k} + \sum_{j \in J} F1_j x_j,$$

subject to Equations 2.5 - 2.10 and Equation 2.16.
Upstream network (CC) model

$$\min \quad \sum_{k \in K} \sum_{l \in L} C4_{kl} u_{kl} + \sum_{l \in L} \sum_{m \in M} C5_{lm} v_{lm} + \sum_{l \in L} F2_l y_l,$$

subject to Equations 2.12- 2.14 and Equation 2.17,

$$R_k = \sum_{l \in L} u_{kl} = \sum_{j' \in J} \left(\sum_{j \in J} \left(\sum_{i \in I} D_i r^*_{ij} \right) S^*_{jj'} \right) t^*_{j'k}, \qquad (2.17)$$

where an asterisk means the optimal value for the downstream problem.
(2) Coordination Between Sub-models

---

[3] A few variant models are solved by taking a volume discount of transport cost and multi-commodity delivery into account [41].

If the optimal values of the coupling quantities, *i.e.*, V_k or R_k, were known *apriori*, we could derive a consistent solution straightforwardly by solving each sub-problem individually. However, since this is not obviously expected, we need to make an adjusting process as follows.

Step 1: For tentative V_k (initially not set forth), solve the downstream problem.

Step 2: After calculating R_k based on the above result, solve the upstream problem.

Step 3: Reevaluate V_k based on the above upstream optimization.

Step 4: Repeat until no more change in V_k has been observed.

In addition, we rewrite the objective function of the downstream problem by relaxing the coupling constraint in terms of the Lagrange multiplier as follows:

$$\sum_{i \in I} \sum_{j \in J} D_i C1_{ij} r_{ij} + \sum_{j \in J} \sum_{j' \in J} \left(\sum_{i \in I} D_i r_{ij} \right) C2_{jj'} s_{jj'} + \sum_{j \in J} F1_j x_j - \sum_{k \in K} \lambda_k V_k$$
$$+ \sum_{j' \in J} \sum_{k \in K} \left(\sum_{j \in J} \left(\sum_{i \in I} D_i r_{ij} \right) s_{jj'} \right) (C3_{j'k} + \lambda_k \rho_k) t_{j'k}. \qquad (2.18)$$

The last term of Equation 2.18 implies that recosting the transport cost $C3_{j'k}$ can conveniently play the role of coordination. It is simply carried out as $C3_{j'k} := C3_{j'k} + \text{constant} \times \rho_k$. From the statements so far, we know that the coordination can be viewed as the auction on the transportation cost so that the procurement becomes most suitable for the entire chain. By virtue of the increase in accuracy by computing V_k and R_k along with the iteration, we can expect convergence from such a coordination.

(3) Procedure for a Coordinated Solution

To reduce the computation load, we further break down each sub-problem into two levels, *i.e.*, the upper level problem to determine the locations and the lower one to determine the routes. Taking such hierarchical approach, we can apply such a hybrid method that will bring about the following advantages:

• In the upper level problem, we can shrink the search space dramatically by confining the search to location only.
• The lower level problem is transformed into a problem that is possible to solve extremely effectively.

As a drawback, we need to solve repeatedly one of the two sub-problems subject to the foregoing result of the other sub-problem in turn. However, the computational load of such an adjustment is revealed to be moderate and effective [27].

Presently, we can solve the upstream problem following the method that applies tabu search [9, 10] for the upper level and mathematical programming for the lower level (hybrid tabu search). Moreover, the lower problem of the upstream network becomes a special type of linear programming referring to the minimum cost flow (MCF) problem [42]. In practice, the original graph representing physical flow (Figure 2.16a) can be transformed into the graph shown in Figure 2.16b, where the label on an arrow and edge indicate cost and capacity, respectively. This transformation is carried out based on the following procedure.

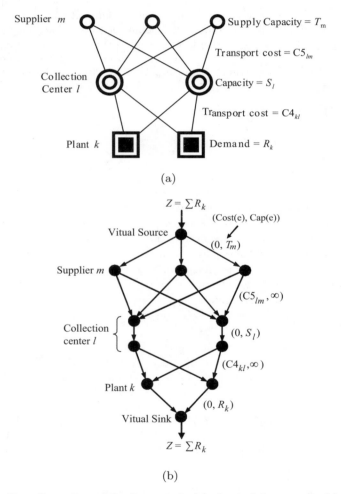

Supplier m 〇 〇 〇 Supply Capacity $= T_m$

Transport cost $= C5_{lm}$

Collection Center l ◎ ◎ Capacity $= S_l$

Transport cost $= C4_{kl}$

Plant k ▣ ▣ Demand $= R_k$

(a)

$Z = \sum R_k$

(Cost(e), Cap(e))

Vitual Source

$(0, T_m)$

Supplier m

$(C5_{lm}, \infty)$

Collection center l {

$(0, S_l)$

$(C4_{kl}, \infty)$

Plant k

$(0, R_k)$

Vitual Sink

$Z = \sum R_k$

(b)

Fig. 2.16. Transformation of the flow graph: (a) physical flow graph, (b) minimum cost flow graph

Step 1: Place the node corresponding to each facility. In particular, double the nodes of hub facilities (CC).

Step 2: Add two imaginary nodes termed source (root of the graph) at the top of the graph and the node termed sink at the bottom of the graph.

Step 3: Connect between nodes with the edge labeled by $(cost(e), capacity(e))$ as follows:

- label the edge between source and supplier by $(0, T_m)$,
- label the edge between supplier and CC by $(C5_{lm}, \infty)$,
- label the edge between the duplicated CC by $(0, S_l)$,
- label the edge between CC and plant by $(C4_{kl}, \infty)$,
- label the edge between plant and sink by $(0, D_k)$.

Step 4: Set the amount of flow $\Sigma_i D_i$ at the source so that the total demand is satisfied.

On the other hand, in the downstream problem, the lower level problem refers to the IP due to the single allocation condition. It is described as the shortest path problem if we neglect the capacity constraints on DCs or Equations 2.10 and 2.12. After all, it is possible to provide another efficient hybrid tabu search that employs the sophisticated Dijkstra method to solve the shortest path problem with the capacity constraints [43]. First, the Lagrange relaxation is used to cope with the capacity constraints. Then the idea simulating an auction on the transport cost is conveniently applied. Thereat, if a certain DC would not satisfy its capacity constraint, we can consider that it occurred due to the too cheap transport costs connectable to that DC. So if we raise such cost, some connections may move on other cheaper routes in the next call. Thus adjusting the transportation cost depending on the violation amount like $\hat{C}1_{ij} := C1_{ij} + \mu \cdot \Delta P_i$, and similarly for $C2$, all constraints are expected to be satisfied at last. Here μ and ΔP_i denote a coefficient related to Lagrange multiplier and the violated amount at the i-th DC.

Finally, the entire procedure is summarized as follows.

Step 1: Set all parameters at their initial values.

Step 2: Under the prescribed parameters, solve the downstream problem by using hybrid tabu search.

2.1: Provide the initial location of DCs.

2.2: Decide on the routes covering the plants, DCs, and customers by solving the capacitated shortest path problem.

2.3: Revise the DCs' location repeatedly until the stopping condition of tabu search is satisfied.

Step 3: Compute the necessary amount of the plant based on the above result.

Step 4: Solve the upstream problem using hybrid tabu search.

4.1: Provide the initial location of CCs.

4.2: Decide on the routes covering the suppliers, CCs and plants from MCF problem.

4.3: Revise the CCs' location according to tabu search.

Step 5: Check the stopping condition. If it is satisfied, stop.
Step 6: Recalculate the transport costs between plants and DCs, and go back
to Step 2.

B. Example of Supply Chain Optimization

The performance of the above method is evaluated by solving a variety of
benchmark problems whose features are summarized in Table 2.1. They are
produced by generating the nodes whose appearance rates become approximately 3: 4: 1: 6: 8 among suppliers, CCs plants, DCs, and customers. Then
the transport cost per unit demand is given by the value corresponding to the
Euclid distance between each node. The demand and capacity are decided
randomly between certain intervals.

Table 2.1. Properties of the benchmark problem

Prob. ID	Sply	CC site	Plant	DC site	Cust	Combination*
b6	84	96	6	108	120	2.6×10^{61}
b7	98	112	7	126	140	4.4×10^{71}
b8	112	128	8	144	160	7.6×10^{81}

* Number of combinations regarding CC and DC locations

In tabu search, we explore the local search space by applying three operations such as add, subtract, and swap with the prescribed probability as
shown in Table 2.2. By letting the attributes of the candidates for neighbor
state be open and closed, we provide the following two rules to prepare a tabu
list with a length of 50.

Rule 1: Prohibit the exchange of attributes when the updated solution can
improve the current solution.
Rule 2: Prohibit keeping the attribute as it is when the updated solution fails.

Table 2.2. Employed neighborhood operations

Type	Probability	Operation
Add	$p_{\mathrm{add}} = 0.1$	Let closed hub v_{ins} open
Subtract	$p_{\mathrm{subtract}}=0.5$	Let opened hub v_{del} close
Swap	$p_{\mathrm{swap}}=0.4$	Let closed hub v_{ins} open and opened hub v_{del} close

   The results summarized in Table 2.3 reveal that the expansion of the computation load of the hybrid tabu search[4] is considerably slow with the increase in problem size compared with commercial software like CPLEX (OPL-Studio) [45]. In Figure 2.17, we present the convergence features including those of downstream and upstream problems. Here, the coordination method works adequately to reduce the total cost by bargaining over the gain at the procurement chain for the loss at the distribution chain. In addition, only a small number of iterations (no more than ten) is required by convergence. By virtue of the generic nature of metaheuristic algorithms, this claims that the converged solution might almost attain the global optimum.

Table 2.3. Performance with commercial software

| Prob. ID | Hybrid tabu search | | OPL-Studio |
	Time [sec]	Appr. rate[*1]	Time [sec] (rate)
b6A	123	1.005	7243 (59)
b6B	78	1.006	15018 (193)
b7A	159	1.006	27548 (173)
b7B	241	1.005	44129 (183)
b8A	231	1.006	24hr[*2](>37)
b8B	376	1.003	24hr[*2](>230)

CPU: 1GHz (Pentium3), RAM: 256MB
[*1] Approximation rate = attained / final sol. of OPL.
[*2] Solution by 24hr computation

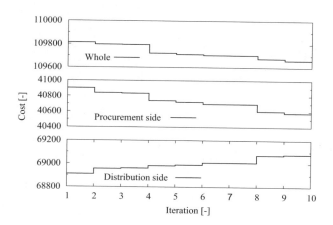

Fig. 2.17. Convergence features along the iteration

---

[4] The MCF problem was solved using a code by Goldberg termed CS2 [44].

2.4.2 Sequencing Planning for a Mixed-model Assembly Line Using SA

For a relevant injection sequencing on a mixed-model assembly line, one of the major aspects is to level out the workload at each workstation against variations of assembly time per product model [46]. Another one is to keep the usage rate of every part constant at the assembly line [47]. These two aspects have been widely discussed in the literature. Usually, to keep production balance and to prevent line stoppage, a large work-in-process (WIP) inventory is required between two lines operated in different production manners, *e.g.*, the mixed-model assembly line and its preceding painting line in the car industry. In other words, achieving these two goals proportionally can bring about a reduction of the WIP inventory. In the following, therefore, we consider a sequencing problem that aims at minimizing the weighted sum of the line stoppage times and the idle time of workers.

A. Model of a Mixed-model Assembly Line with a Painting Line

Figure 2.18 shows a mixed-model assembly line including a painting line where each product is supplied from the foregoing body line every cycle time (CT). The painting line is composed of sub-painting, main painting and check processes. Re-painting repeats the main painting twice to correct defective products. The defective products are put in the buffer after correction. From the buffer, necessary amounts of product are taken out in order of the injection sequence at the mixed-model assembly line. It is equipped with K workstations on a conveyor moving at constant speed. At each workstation, a worker assembles the prescribed parts into the product models.

Furthermore, we assume the following conditions.

1. Paint defects occur at random.
2. The correction time of defective product varies randomly.
3. The production lead-time of the painting line is longer than that of the assembly line.

The sequencing problem under consideration is formulated as follows:

$$\min_{\pi \in \Pi} \ \rho_p \times B^t + \rho_a \times \sum_{t=1}^{T} \max_{1 \leq k \leq K} (P_k^t, A_k^t) + \rho_w \times \sum_{t=1}^{T} \sum_{k=1}^{K} W_k^t,$$

subject to

$$\sum_{i=1}^{I} z_i^t = 1, \quad t = 1, \ldots, T, \tag{2.19}$$

$$\sum_{t=1}^{T} z_i^t = d_i, \quad i = 1, \ldots, I, \tag{2.20}$$

where the notation is as follows.

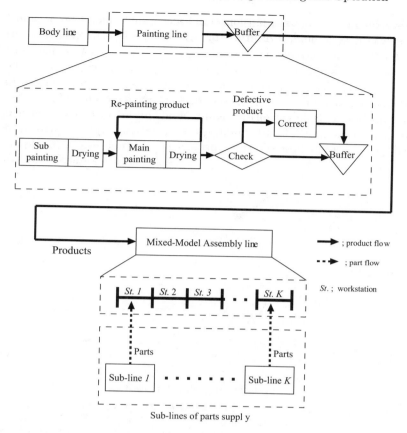

Fig. 2.18. Scheme of a mixed-model assembly line and a painting line model

I: number of product models.

K: number of workstations.

T: number of injection periods.

π: injection sequence over a planning horizon (decided from z_i^t).

Π: set of sequences ($\pi \in \Pi$).

B^t: line stoppage time due to product shortage at injection period t.

P_k^t: line stoppage time due to part shortage at workstation k at injection period t.

A_k^t: line stoppage time by work delay of a worker. This happens when the workload exceeds CT in workstation k at injection period t.

W_k^t: idle time of worker at workstation k at injection period t.

z_i^t: 0-1 variable that takes 1 if the product model i is supplied to the assembly
line at injection period t. Otherwise, 0.

d_i: demand of product model i over T.

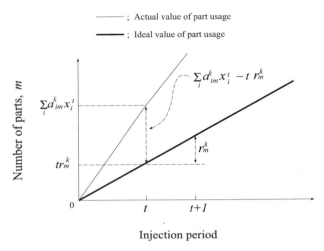

        ————— ; Actual value of part usage

        ▬▬▬ ; Ideal value of part usage

Injection period

Fig. 2.19. Line stoppage time based on the goal chasing method [48]

We suppose that the objective function is described by a weighted sum of
the line stoppage times and the idle time, where ρ_p, ρ_a and ρ_w are weighting
factors ($0 < \rho_p, \rho_a, \rho_w < 1$). Among the constraints, Equation 2.19 indicates
that plural products cannot be supplied simultaneously, and Equation 2.20
requires that the demand of each product model be satisfied.

Figure 2.19 illustrates a situation where the part shortage occurs at the
workstation k when the quantity of part m used ($\sum_i a_{im}^k x_i^t$) exceeds its ideal
quantity (tr_m^k) at the injection period t. Then, P_k^t is given as follows:

$$P_k^t = \max[\max_{1 \le m \le M} (\frac{\sum_{i=1}^I a_{im}^k x_i^t - tr_m^k}{r_m^k} \mathrm{CT}), \ 0],$$

where a_{im}^k is the quantity of part m required for model i, x_i^t the accumulative
amount of production for model i during injection period from 1 to t, $i.e.$,

$$x_i^t = \sum_{l=1}^t z_i^l, \quad (i = 1, \dots, I). \tag{2.21}$$

Moreover, r_m^k denotes the ideal usage rate of part m, and M the maximum
number of parts used on the workstation.

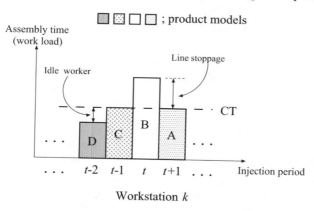

Fig. 2.20. Line stoppage due to workload unbalance

On the other hand, Figure 2.20 show a simple example of how line stoppage or idle work occurs due to variations of workloads. Each product model with different workloads are put into workstation k along injection period. Since the assembly time (workload) exceeds CT at injection period t, the line stoppage occurs whereas idle work occurs at $t - 2$. By knowing these, the line stoppage time A_k^t and the idle time W_k^t can be calculated from Equations 2.22 and Equation 2.23, respectively,

$$A_k^t = \max(L_k^t - \mathrm{CT},\ 0), \tag{2.22}$$

$$W_k^t = \max(\mathrm{CT} - L_k^t,\ 0), \tag{2.23}$$

where L_k^t denotes the working time of a worker at workstation k at injection period t.

Noticing that the product models from the painting line can be viewed equivalently as the parts from a sub-line in the mixed-model assembly line, we can give the line stoppage time B^t due to part shortages as Equation 2.24,

$$B^t = \max(\frac{x_i^t - tr_{pi}}{r_{pi}}\mathrm{CT},\ 0),\ t = 1, \dots, T,\ i = 1, \dots, I, \tag{2.24}$$

where r_{pi} is the supply rate of product model i from the painting line over the entire injection periods. Consequently, Equation 2.24 shows the time difference between the actual injection time of the product model i and the ideal one. Here we give r_{pi} like Equation 2.25 by taking the correction time of defective products at the painting line into account,

$$r_{pi} = \frac{d_i}{T + [\sigma d_i C_i]}, \qquad i = 1, \ldots, I, \qquad (2.25)$$

where σ is the defective rate of products at the painting line, C_i the correction time for the defective product model i, and $[\cdot]$ is a Gauss symbol.

Furthermore, to improve the above prediction, r_{pi} is revised at every production period $(n = 1, \ldots, N)$ according to the following procedures.

Step 1: Forecast r_{pi} from the input order to the painting line at $n = 1$ (see Figure 2.21a).

Step 2: After the injection at production period n is completed, obtain the quantity and the completion time (called "delivery-information" hereinafter) of product model i in the buffer.

Step 3: Update r_{pi} based on the delivery information of model i acquired at $n - 1$ (see Figure 2.21b).

    3-1: Generate the supply rate F_{ij} $(j = 1, 2, \ldots)$ at every injection period when product model i is put into the buffer.

    3-2: Average F_{ij} and r_{pi} to obtain the supply rate r'_{pi} of the product model i at n.

Step4: If $n = N$, stop. Otherwise, Let $n := n + 1$ and go back to Step 2.

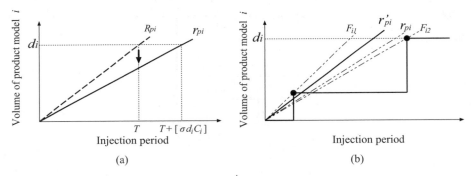

Fig. 2.21. Forecast scheme of r_{pi} and r'_{pi}: (a) estimation of r_{pi} $(n = 1)$, (b) re-evaluation of r_{pi} $(n > 1)$

B. An Example of a Mixed-model Assembly Line

Numerical experiments are carried out under the conditions shown in Table 2.4. Weighting factors ρ_p, ρ_a and ρ_w are set as 0.5, 0.4 and 0.1, respectively. Moreover, the results are evaluated based on the average over 100 data sets generated randomly. To cope with the sequencing problem that belongs to a NP-hard solution procedure, SA is applied as a solution method for deriving a near optimal solution. We give a reference state by the random sequence of injection so as to satisfy Equations 2.19 and 2.20. Then swapping two arbitrarily chosen product models in the sequence generates the neighbors of state. In the exponential cooling schedule, the temperature decreases by a fixed factor 0.8 at each step from the initial temperature 100 to the end during 150 iterations.

Table 2.4. Input parameters

Cycle time, CT [min]	5
Station number, K	100
Product model, I	10
Total production number, $\sum_i d_i$	100
Injection period, T	100
Production period, N	30
Defective rate	0.2
Correction time [min]	[15, 25]

The advantages of the total optimization ("Total sequencing") were compared with the result obtained when neglecting the two terms in the objective function, *i.e.*, $\rho_p = \rho_w = 0$ ("Level sequencing").

Table 2.5. Comparison of sequencing strategies

	WIP inventory volume	Line stoppage time [min]	Idle time [min]
Total sequencing	28.7	43.7	4.2
Level sequencing	37.5	31.2	4.1

In Table 2.5, the WIP inventory volume means the value necessary for preventing line stoppage due to product shortage while the line stoppage time and idle time are the times incurred by the non-leveling of the parts usage and the workloads at the assembly line, respectively. Though the WIP inventory of "Total sequencing" is smaller than that of the "Level sequencing", the line stoppage and the idle times are a little inferior to the previous result. Therefore, the advantage of the optimization actually refers to the relevant management of the WIP inventory between two lines. As illustrated in Figure 2.22, "Total sequencing" is known to achieve the drastic decrease and stable volume in the inventory compared with "Level sequencing".

2.4.3 General Scheduling Considering Human–Machine Cooperation

A number of resources controlled by computers are now popular in manufacturing *e.g.*, CNC machine tools, robots, AGVs, and automated warehouses.

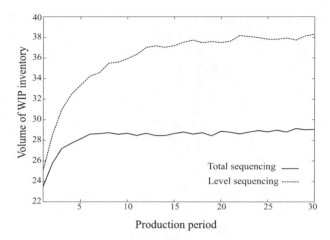

Fig. 2.22. Features of the WIP inventory along a production period

There, the role of computers is to execute the prescribed tasks automatically according to the production plans. Therefore, the advanced production resources automated by the computer are expected to explore the next generation of manufacturing systems [49]. In the near future, autonomous machine tools and robots might produce various products in flexible manners. In the current systems, however, the role of the human operator is still important. In many factories, multi-skilled operators manipulate the multiple machine tools while moving among the multiple resources. Such a situation makes it meaningless to ignore the role of operators and make a plan confined only to the status of non-human resources.

This point of view requires us to generalize the scheduling problem associated with the cooperation between human operators and resources [37]. Based on the relationship between the resources assigned to the job, incidental operations such as loading and unloading of the products are analyzed according to material flows. Then, a modified dispatching rule is applied to solve the scheduling problem.

A. Operation Classes for Generating a Schedule

The following notations will be used since production is related to a number of jobs, operations and processes associated with the job. Moreover, the term "process" will be used when we emphasize dealing with a product while "operation" will be used when we represent the manipulation of resources.

$j_{\eta,i}^{\zeta,v}$: v-th operation processed by resource ζ and i-th process for product η regarding parameter j.

s: starting time of the job.

f: finishing time of the job.
p: processing time of the job.

The scheduling problem is usually formulated under the following assumptions.

1. Every resource can perform only one job at a time.
2. Every resource can start an operation after a preceding process has been finished.
3. The processing order and the processing time are given, and any change of the processing order is prohibited.

Under these conditions, the scheduling is to determine the operating order assigned to each resource. Figure 2.23 illustrates the Gantt charts for two possible situations of a job processed by machines ξ and ζ. As shown in Figure 2.23a, it is possible to start the target operation of resource ζ immediately after the preceding operation has been finished. In contrast, as shown in Figure 2.23b, since machine ζ can perform only one job at a time, resource ζ cannot begin to process even if resource ξ has finished the preceding process.

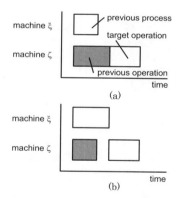

Fig. 2.23. Dependency of jobs processed by two machines: (a) on the previous operation, (b) on the previous process

Therefore, the starting time of the target job can be determined as follows:

$$s_{\eta,i}^{\zeta,v} = \max[f^{\zeta,v-1}, f_{\eta,i-1}], \qquad (2.26)$$

where operator $\max[\cdot]$ returns the greatest value of the arguments. On the other hand, the finishing time is calculated by the following equation:

$$f_{\eta,i}^{\zeta,v} = s_{\eta,i}^{\zeta,v} + p_{\eta,i}^{\zeta,v}.$$

In addition, we need to consider the following aspects for the generalization of scheduling. In the conventional scheduling problem, it is assumed that each resource receives one job from another resource, then processes it and transfers it to another resource. However, in real world manufacturing, multiple resources are commonly employed to process a job. Figure 2.24 shows three types of Gantt charts for cases where multiple resources are used for manufacturing.

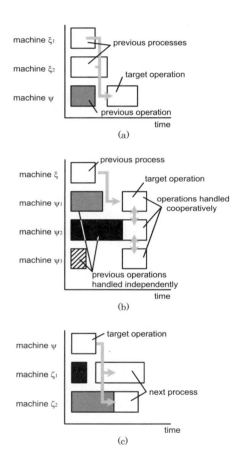

Fig. 2.24. Classification of a schedule based on material flows: (a) parts supplied from multiple machines, (b) operations handled cooperatively by multiple machines, and (c) operations of plural jobs using parts supplied from one machine

In the first case (a), one resource receives one job from the multiple resources. This type of material flow, called "merge", corresponds to the case where a robot assembles multiple parts supplied to it from the multiple resources, for example. In the second case (b), multiple resources are assigned to a job. However, each resource cannot begin to process the job until all resources have finished the preceding jobs. This type of production is known as "cooperation". Examples of the cooperation are cases where an operator manipulates a machine tool, and where a handling robot transfers a job from AGV to machine tool. The last one (c) corresponds to "distribution", which is the case where several resources receive the job individually from another resource. Carrying several types of parts by truck from a subcontractor is a typical example of this case. Various resources cannot begin to process until all trucks arrive at the factory. In these cases, the starting time of the target job is determined as follows.

$$s_{\eta,i}^{\psi_\alpha,v} = \max[\{f_{\eta,i-1}^{\xi_\gamma,v-1}\}, \ \{f^{\psi_\beta,w-1}\}, \ f^{\psi_\alpha,v-1}],$$

where ξ_γ is every resource processing the preceding process of the job $j_{\eta,i}^{\psi_\alpha,v}$ and ψ_β every resource processing the job cooperatively with resource ψ_α. Resource ψ_β processes the job $j_{\eta,i-1}$ as the w-th operation, and $\{\cdot\}$ shows a set of finishing times f.

Jobs like loading and unloading are respectively considered as a pre-operation and a post-operation incidental to the main job (incidental operation). Status check and execution of NC program by a human operator are alos viewed as such operations. In conventional scheduling, these jobs are likely to be ignored because they take a much shorter time compared with the main job. However, the role of these operations are still essential whenever their processed times are insignificant. For example, the resources cannot begin the process without a safety check by a human operator even in current automated manufacturing.

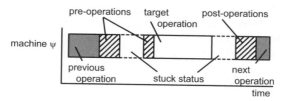

Fig. 2.25. Pre-operation and post-operation

Figure 2.25 illustrates the case where multiple pre-operations and post-operations are related to the main job (noted as the target operation). Between the two incidental operations and/or between the incidental operation and the main job, there arises an undesirable idle time or stuck time during which the resource cannot execute the other job. For generalizing the scheduling, concerns with these operations are also unavoidable.

B. Solution Method

Generally speaking, an appropriate dispatching rule can derive a practical schedule even for the real world problem with a large number of products and resources. To deal with the complicated situations mentioned above in a practical manner, it makes sense to apply this kind of knowledge or an empirical optimization method. A modified earliest start time (EST) rule is effective for obtaining a schedule to level out the waiting times. It is employed as follows.

Step 1: Make an executable job list $\{j_{\eta,i}^{\zeta,v}\}$ where job $j_{\eta,i}^{\zeta,v}$ is the first job of the product or the preceding job $j_{\eta,i-1}$ assigned on the schedule.

Step 2: Calculate the starting time $s_{\eta,i}^{\zeta,v}$ of the job $j_{\eta,i}^{\zeta,v}$ by Equation 2.26. If the operator manipulating machine ζ for processing job $j_{\eta,i}^{\zeta,v}$ engaged in the manipulation of another machine ξ before $j_{\eta,i}^{\zeta,v}$, then modify $s_{\eta,i}^{\zeta,v}$ using the following equation:

$$\hat{s}_{\eta,i}^{\zeta,v} = s_{\eta,i}^{\zeta,v} + t_{\xi,\zeta},$$

where $t_{\xi,\zeta}$ is the moving time of the operator from machine ξ to machine ζ.

Step 3: Select the job that can begin the process earliest. If there are plural candidates, select the job that has the most work to do.

Step 4: Repeat from Step 1 through 3 until all jobs are assigned to the resources.

C. Examples of a Schedule with a Human Operator

To illustrate the validity of the above discussions, a job shop scheduling problem is solved under the following conditions. Two multi-skilled operators and eight machine tools produce ten products. Both operators can manipulate multiple machine tools. Every job processed by the machine tool requires pre-operation and post-operation by the human operators. These incidental jobs are also identified as the jobs that need cooperation between human operators and machines.

Figure 2.26 shows a Gantt chart partially extracted from the scheduling obtained here. As shown in these figures, one operator manipulates the machine both at the beginning and at the end of jobs. Figure 2.26b shows the case where the moving time of an operator between two machines is short and the operator can move to machine ζ immediately after loading on machine ξ. Staying at machine ζ until the unloading of job B, the operator can return to machine ξ without any delay for unloading job A.

On the other hand, Figure 2.26c shows the case where the operator takes double time to move between these two machines. However, the operating order is the same as before, the stuck time occurs on machine ξ due to the late arrival of the operator.

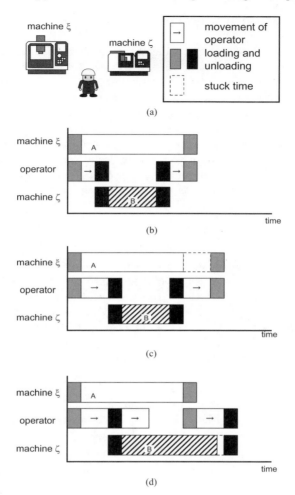

Fig. 2.26. Examples of scheduling with a human operator: (a) operator and machine tools, (b) schedule with loading and unloading by an operator, (c) schedule when an operator takes double time for movement between machine tools, and (d) schedule when job B takes double time for operation

Moreover, Figure 2.26d shows the influence of the job processing time. If the processing time of job B is double, it wastes much time because the operator will not stay at machine ζ. The operator returns to machine ξ immediately after setting up the job on machine ζ and waits for job A to be completed by machine ξ. The stuck time occurring on machine ζ becomes shorter compared with the stuck time occurring on machine ξ if the operator stays at machine ζ. This example clearly reveals the importance of the contribution of operators for a practical schedule.

2.5 Optimization under Uncertainty

There exist more or less uncertain factors in mathematical models employed for manufacturing optimization. As the lead-time for system development, planning and design become longer, systems will suffer unexpected deviations more often and more seriously. However, since it is impossible to forecast every unknown or uncertain factor beforehand, we need to analyze in advance the influence of such uncertainties on state and performance before optimization. Without considering various uncertainties involved in the system model, it may happen that the optimum solution is useful only in the specific situation, or at worst becomes insignificant. Especially when engaging in the real world problems, such an understanding is of special importance to guarantee a certain security, confidence, and economical merit.

There are known several types of uncertainty, associated with the optimization problems, *i.e.*, parameter deviations involved in the objective function and constraints; structural errors of the system model, *e.g.*, linear/non-linear, missing/redundant variables and/or constraints, *etc.* Regarding the nature of uncertain parameters, they are also classified into categories, *i.e.*, deterministic, stochastic and fuzzy deviations. To cope with the uncertainties associated with the optimization problem either explicitly or inexplicitly, much research has been carried out for many years. They refer to technical terms such as sensitivity, flexibility, robustness, and so on. Stochastic optimization, chance constrained optimization and fuzzy optimization are popularly known classes of optimization problems associated with uncertainties.

Leaving the introduction of these approaches to other literature [50], a new interest related to the recent development of metaheuristic optimization methods will be considered here. Deriving an insensitive solution against uncertainties is a major interest in this section.

2.5.1 A GA to Derive an Insensitive Solution against Uncertain Parameters

It is desirable to make the optimal solution adapt dynamically according to the deviation of parameters and/or changes of the environment. For various reasons, however, such a dynamic adaptability is not easy to achieve. Instead, we might take a proper precaution and try to obtain a solution that is robust against the changes. For this purpose, such a problem is often formulated as a stochastic optimization problem that will maximize the expectation of the objective function with uncertain parameters. Similarly, we introduce a few GA methods where fitness is calculated by stochastic parameters like expectation and variance of the objective function. Though GA has been applied to many deterministic optimizations, not so many studies have been carried out on the uncertainties [51, 52, 53, 54]. However, by virtue of the population-based search method through natural selection, GA has a high potential ability to cope with the uncertainties.

First, let us consider the deterministic optimization problem described as follows:

[*Problem*] $\min f(x)$ subject to $x \in X \subseteq \mathrm{R}^n$,

where x denotes a decision variable vector and X its admissible region. Moreover, f is an objective function. On the other hand, the optimization problem under uncertainty is given by

[*Problem*] $\min \ F_w(f(x, w))$ subject to $\begin{cases} x \in X \subseteq \mathrm{R}^n \\ w \in W \subseteq \mathrm{R}^m \end{cases}$.

Since GA popularly handles constraints with the penalty function method, below the uncertainties are assumed to be involved only in the objective function without loss of generality. Moreover, if the influence from uncertainties is evaluated through expectation, the above problem can be re-described as follows:

[*Problem*] $\min \ E_w[f(x, w)]$ subject to $x \in X \subseteq \mathrm{R}^n$,

where $E_w[\cdot]$ denotes the expectation with respect to w. When the probabilistic distribution function $\varphi(w)$ is given, it is calculated by the following equation:

$$E_w = \int_{-\infty}^{\infty} \varphi(w) f(x, w) \mathrm{d}w.$$

On the other hand, when the uncertain parameters deviate randomly within a certain interval, or the probabilistic distribution function is not given explicitly, the above computation is substituted by the average over K samples. In this case, a large number of samples can increase the accuracy of such a computation,

$$E_w = \frac{1}{K} \sum_{i=1}^{K} f(x, w_i).$$

Due to the generic property compared to the natural selection, in GA, individuals with higher adaptability can survive to the next generation even in an environment suffering from (parameter) deviations. This means that these survivors have been exposed to various parameter deviations during all generations long. Accordingly, the solutions obtained there are to be selected based on the expectation computed through a large number of sampling eventually or the most precise evaluation. In other words, GA can concern the uncertain problem altogether and all over the generation as well. Noting the high computational load of GA, however, how to reduce the additional load consumed for such a computation becomes a major point in developing effective methods.

The first method applies the usual GA by simply calculating the fitness from the expectation in terms of the sufficient number of samples in every generation, $i.e.$, $F_i = E_w[\cdot]$. As easily supposed, a very large number of samples is to be evaluated by the end of the search. Usually, the same stopping condition is adopted as same as in the usual GA.

Since the dominant individuals are to be evaluated repeatedly over the generation, it is possible to reduce the load necessary for the correct evaluation of expectation if the inherited information is available. Based on such prospects, the second method [49] uses Equation 2.27 for the calculation of fitness (for simplicity, the following equations are described assuming decision variable is scalar):

$$F_i = \frac{Age_i - 1)H(P_i) + f(x_i, w_j)}{Age_i}, \tag{2.27}$$

where F_i is the fitness of the i-th chromosome, $H(P_i)$ the fitness of one of the parent being closer to each offspring in the search space (its distance is denoted by D). Age_i corresponds to the individual's age that increases with the generation by one, but is reset every generation with the probability $1 - p(D)$. Here, $p(D)$ is given as

$$p(D) = \exp(-\frac{D^2}{\alpha}),$$

where α is a constant adjusting the degree of inheritance. As α becomes larger, it is more likely to inherit the character from the parent and $vice$ $versa$. Since the sampling is limited to only one, this method weighs the contribution of the inheritance based on insufficient information too much on the evaluation of fitness. The individual with the highest age is chosen as the converged solution.

To compromise the foregoing two methods, the third method [56] illustrated in Figure 2.27 takes multiple samplings that are not so large but not only one. They are used to calculate not only the expectation but also the variance. The additional information from the variance can compensate the insufficiency of the inherited information available at the present generation in Equation 2.27. Eventually, the fitness of the i-th individual is given by the following equation:

$$F_i = \frac{(Age_i - 1)H(P_i) - h(\bar{f}_i, \sigma_i^2)}{Age_i},$$

where $h(\bar{f}_i, \sigma_i^2)$ is given by

$$h(\bar{f}_i, \sigma_i^2) = \lambda \bar{f}_i + (1 - \lambda)\sigma_i^2,$$

where λ is a weighting factor and $\bar{f}_i$ and σ_i^2 denote the values of average and standard deviations, respectively,

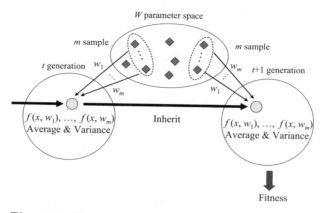

Fig. 2.27. Computation method of fitness by method 3

$$\bar{f}_i = \frac{1}{m} \sum_{j=1}^{m} f(x_i, w_j),$$

$$\sigma_i^2 = \frac{1}{m-1} \sum_{j=1}^{m} (f(x_i, w_j) - \bar{f}_i)^2,$$

where m is the sampling number.

After the stopping condition has been satisfied, the individual with the highest age is chosen as the final solution.

The first test problem to examine the performance of each method is given by the maximization of a two-peaked objective function shown in Figure 2.28.

$$f_1(x, w) = \begin{cases} A_L \sin\{B_L(x + w)\}, & (x + w \in D_L) \\ A_R \sin\{B_R(x + w - \frac{1}{11})\}, & (x + w \in D_R) \end{cases},$$

where $D_L = \{x | 0 \leq x \leq 1/11\}$, $D_R = \{x | \frac{1}{11} \leq x \leq 1\}$. A noisy parameter w deviates in two ways:

1. randomly within $[-0.004, 0.004]$
2. under the normal distribution $N[0, \sigma^2]$.

Furthermore, in the second case, two sizes of deviation are considered, *i.e.*, $\sigma = 0.01$, and 0.05. As known from Figure 2.28, the optimal solution for each σ becomes $x_L = 0.046$ and $x_R = 0.546$, respectively. Table 2.6 compares the results obtained under the condition that the population size = 100, crossover the rate = 0.6, and the mutation rate = 0.02. After the same prescribed computation time (30 s), the final solution is chosen according to the stopping condition of each method.

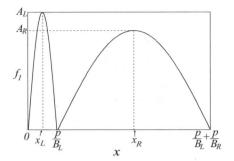

Fig. 2.28. Two-peak problem $f_1(x, w)$, ($A_L = 10$, $A_R = 8$, $B_L = 11\pi$, $B_R = 11\pi/10$, $w = 0$)

Table 2.6. Comparison of numerical results

σ	Method	Solution	Error (%)	m	Generation
0.01	1	0.0436	4.2	20	3000
($x_L = 0.046$)	2	0.0486	22.2	1	12000
	3	0.0458	2.3	5	8000
0.05	1	0.539	2.0	20	3000
($x_R = 0.546$)	2	0.523	9.1	1	12000
	3	0.545	1.7	5	8000

In every case, the third method outperforms the others. On the other hand, all results of the case $\sigma = 0.01$ are inferior to those of $\sigma = 0.05$, since around the optimal solution for $\sigma = 0.01$ (x_L), the sensitivity of f_1 with w is higher than that of the optimal solution for $\sigma = 0.05$ (x_R).

Another test problem with the five-modal objective function shown in Figure 2.29 is also solved by each method,

$$f_2(x, w) = \begin{cases} a(x, w)|\sin(5\pi(x + w))|^{0.5}, & (0.4 < x + w \le 0.6) \\ a(x, w)\sin^6(5\pi(x + w)), & \text{otherwise} \end{cases},$$

where $a(x, w) = \exp[-2\ln 2(\frac{(x+w)-0.1}{0.8})^{0.2}]$.

In this problem, the noisy parameter deviates under the normal distribution with $\sigma = 0.02$ and 0.04. As shown in Figure 2.29, the optimal solution for each deviation locates at $x_L = 0.1$ and at $x_R = 0.492$, respectively. Figure 2.30 shows the behavior of the tentative solution during the generation for $\sigma = 0.02$. From this, it is known that the third method attains the optimal solution x_L fast, and keeps it steadily. This means that the result will not be affected by the wrong selection of the stopping condition, or the oldest individual can dwell on the optimal state safely. On the other hand, the second method is inferior to the others. Figure 2.31 describes the result for $\sigma = 0.04$.

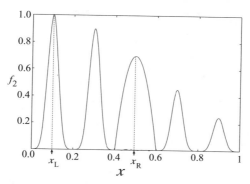

Fig. 2.29. Five-peak problem $f_2(x, w), (w = 0)$

In this case, the third method also outperforms the others. These results claim that the third method can derive the solution steadily and safely regardless of the stopping conditions.

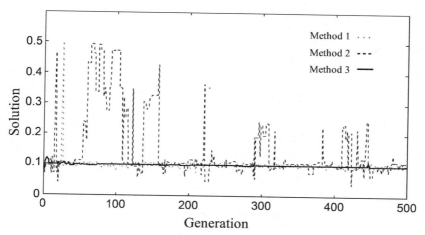

Fig. 2.30. Convergence property ($\sigma = 0.02$)

2.5.2 Flexible Logistic Network Design Optimization

Under the influence of globalization and the introduction of advanced transportation systems, industrial markets are acknowledging the importance of flexible logistic systems favoring just-in-time and agile manufacturing. Focusing on the logistic systems associated with supply chain management (SCM), a method termed hybrid tabu search is applied to solve the problem under deterministic customer demand [43]. In reality, however, a precise forecast

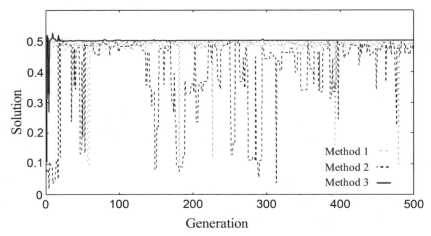

Fig. 2.31. Convergence property ($\sigma = 0.04$)

of demand is quite difficult. An incorrect estimate causes either insufficient production when forecast goes below the actual demand or undue expenditure due to large inventory. It is important, therefore, to formulate the problem by taking into account uncertainty in the demand. In fact, by assuming certain stochastic deviation, two-stage formulations using stochastic programming have been studied [57, 58]. However, these approaches seem to be ineffective for designing a flexible logistic network for the following two reasons. First, customer satisfaction is evaluated by the demand basis but it is left unrelated to other important factors like cost, flexibility, *etc.* Second, they are unconscious of taking a property of decision variables into account whether they are soft (control) or hard (design) variables.

To show an approach for deriving a flexible network against uncertain demands, let us consider a hierarchical logistic network as depicted in Figure 2.32, and define index sets I, J and K for customer (CT), distribution center (DC) and plant (PL), respectively. It is assumed that customer i has an uncertain demand D_i obeying a normal distribution. To consider this problem, a fill rate of demand termed service level is defined as follows:

$$s\left(\alpha\sigma\right) = \int_{-\infty}^{\alpha\sigma} N\left[p_0,\ \sigma\right] \mathrm{d}p \quad \left(\alpha : \mathrm{naturalnumber}\right), \tag{2.28}$$

where $N[\cdot]$ stands for the normal distribution with average p_0 and standard deviation σ. The service level corresponds to the probability that the network can deliver products to customers whatever deviation of the demand might occur within the prescribed extent.

For example, the network designed for the average demand can present 50% service level, and 84.13% for the demand corresponding to $p_0 + \sigma$. Now the problem is to minimize the total transportation cost with respect to the lo-

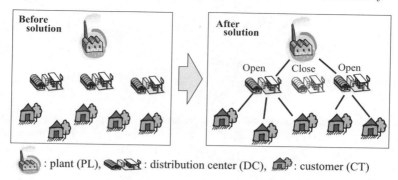

: plant (PL), ⬛⬛ : distribution center (DC), ⬛ : customer (CT)

Fig. 2.32. Scheme of a logistic network

cation of DC and the selection of a route between the facilities while satisfying the service level. The following development also assumes the following:

1. Every customer is supplied via a route only as from PL to DC and from DC to CT.
2. To avoid a separate delivery, each connection is limited to only one linkage (single allocation).

Now, the problem without taking the demand deviation into account is given by the following mixed 0-1 programs [40], which is a variant formulation[5] of the downstream problem of logistic optimization in Sect. 2.4.1:

$$[Problem] \quad \min \sum_i \sum_j f_{ij} E_{ij} + \sum_j \sum_k g_{jk} G_{jk}, \qquad (2.29)$$

subject to

$$\sum_j y_{ij} = 1, \quad \forall i \in I, \qquad (2.30)$$

$$f_{ij} \geq y_{ij} D_i, \quad \forall i \in I, \quad \forall j \in J, \qquad (2.31)$$

$$\sum_i f_{ij} \leq x_j U_j, \quad \forall j \in J, \qquad (2.32)$$

$$x_j = \sum_k z_{jk} M, \quad \forall j \in J, \qquad (2.33)$$

$$g_{jk} \leq z_{jk} M, \quad \forall j \in J, \quad \forall k \in K, \qquad (2.34)$$

$$\sum_k g_{jk} = \sum_{ij} f_{ij}, \quad \forall j \in J, \qquad (2.35)$$

---

[5] Fixed charge of location is ignored. Instead, the number of locations is set at p and delivery between DC and DC is prohibited in this model.

$$\sum_j g_{jk} \leq S_k, \quad \forall k \in K, \tag{2.36}$$

$$\sum_j x_j = p, \tag{2.37}$$

$$f, g : \text{integer}, x, y, z \in \{0, 1\},$$

where x_j denotes a binary variable that takes 1 when DC opens at the j-th candidate and 0 otherwise. The binary variables y_{ij} and z_{jk} represent the status of connection between CT and DC, and DC and PL, respectively. These two binary variables (y_{ij} and z_{jk}) become 1 when connected and 0 otherwise. Quantities f_{ij} and g_{jk} are shipping amounts from DC to CT, and from PL to DC, respectively.

The objective function stands for the total transportation cost where E_{ij} denotes unit transportation cost between the i-th CT and the j-th DC and G_{jk} that between the j-th DC and the k-th PL.

On the other hand, each constraint denotes the conditions as follows: Equation 2.31 denotes demand satisfaction where D_i represents the i-th demand; Equations 2.30 and 2.33 the single linkage conditions; Equations 2.32 and 2.36 capacity constraints where U_j is capacity at the j-th DC and S_k that at the k-th PL; Equation 2.35 flow balance; Equation 2.37 the required number of open DC. Moreover, M in Equations 2.33 and 2.34 represents a very large number.

To consider the problem, the decision variables are classified into hard and soft variables depending on their generic natures. Hard variables are not allowed to change once they have been determined (*e.g.*, DC location). On the other hand, soft variables can change according to the demand deviation (*e.g.*, distribution route). Then a two-level problem is formulated based on the considerations from flexibility analysis [60] as follows:

$$[Problem] \quad \min_{x,u,w} C_T(x, u, w | p_0),$$

$$\text{subject to}$$

$$(x, u, w) \in F(x, u, w | p_0), \tag{2.38}$$

$$||u - v|| \leq 2\xi, \tag{2.39}$$

$$\min_{x,v,w'} C_T(x, v, w' | p_r),$$

$$\text{subject to} \quad (x, v, w') \in F(x, v, w' | p_r), \tag{2.40}$$

$$x, u, v \in \{0, 1\}, \quad w, w' : \text{integer},$$

where x denotes the location of DC (hard variable), u and v correspond to the soft variables denoting the route for the nominal (average) demands, and the deviated demands, respectively. When $|| \cdot ||$ denote the Hamming distance, ξ refers to the allowable number of route changes. This is equivalently described as Equation 2.39. Moreover, w and w' represent the other variables in the original problem at the nominal and the deviated states, respectively.

Also, $C_T(\cdot|p_0)$ and $F(\cdot|p_0)$ in Equation 2.38 symbolically express the objective function (Equation 2.29) and the feasible region at the nominal (Equations 2.30 through 2.37), respectively. Similarly, Problem 2.40 stands for the optimization at the deviated state. Due to the linearity of the constraints regarding demand satisfaction, $i.e.$, Equation 2.31, we can easily describe the permanently feasible region [61, 62]. This condition guarantees the feasibility even in the worst case of parameter deviations regardless of the design and control adopted. Accordingly, the demand D_i in $F(\cdot|p_r)$ must be replaced with the value corresponding to the prescribed service level. Finally, the lower level problem tries to search the optimal route while satisfying the feasibility against every deviation under the DC location decided at the upper level problem.

Even in the case where uncertainties are not considered, the formulated problem belongs to the class of NP-hard problems. It becomes especially difficult to obtain a rigid optimal solution mathematically as the problem size expands. The hybrid tabu search is applied as a core method to solve this problem repeatedly for a parametric study regarding ξ. It is necessary to engage in a tradeoff analysis on the flexible logistics decision at the next stage.

The effectiveness of the approach is examined through a variety of problems where the number of customers ranges from 50 to 150. Moreover, the number of plants $|K|$, candidate DC $|J|$, designated open DC p and customer $|I|$ are set at the ratio 5: 15 : 7: 50, and these facilities are located randomly. Then unit transportation costs E_{ij} and G_{jk} are given to be proportional to the Euclid distance between them.

Three benchmark problems are solved to examine the properties of the flexible solution through comparison with other methods. Table 2.7 shows the results of the three strategies, $i.e.$, the flexible decision (F-opt.), nominal one (N-opt.), and conservative one (W-opt.). N-opt. and W-opt. are derived from the other optimizations described below, respectively,

$$\min C_T(x, u, w|p_0) \quad \text{subject to} \quad (x, u, w) \in F(x, u, w|p_0),$$

$$\min C_T(x, v, w'|p_r) \quad \text{subject to} \quad (x, v, w') \in F(x, v, w'|p_r).$$

Then, the objective values are compared with each other both at the nominal (p_o) and the worst $(p_o + 3\sigma)$ states when $\xi = 5$. The values in parenthesis express the rates to the respective optimal values. In every case, N-opt. is unable to cope with the deviated state. On the other hand, though W-opt. has an advantage at the worst state, its performance degrades outstandingly at the nominal state. In contrast, F-opt. can present better performance in the nominal state while keeping a nearly optimal value in the worst case.

Results obtained from another class of problems reveal that the more difficult the decision environment and the more seriously the deviated situation become, the more the flexible design takes the advantage.

Table 2.7. Comparison of results for the benchmark problem

Problem ID $(D\text{-}|K|\text{-}|J|(p)\text{-}|I|)$	Strategy	At nominal state	At worst state
	F-opt.	45846 (1.25)	77938 (1.04)
D-5-15(7)-50	N-opt.	36775 (1.00)	NA
	W-opt.	58377 (1.59)	74850 (1.00)
	F-opt.	38127 (1.03)	47661(1.04)
D-10-30(14)-100	N-opt.	36918 (1.00)	NA
	W-opt.	39321 (1.06)	45854 (1.00)
	F-opt.	40886 (1.07)	48244 (1.05)
D-15-45(21)-150	N-opt.	38212 (1.00)	NA
	W-opt.	45834 (1.19)	45899 (1.00)

To make a final decision associated with the flexibility, the dependence of adjusting margin ξ on the system performance or total cost needs to be examined. Since certain amounts of margin (ξ) can reduce the degradation of performance (total cost) effectively, we can derive a rational decision by compromising the attainability of these factors. An example of the tradeoff analysis is shown in Figure 2.33. Due to the tradeoff between the total cost and ξ, which increases along with the amount of deviation, decision making at the next step should be addressed in terms of the discussion about the sufficient service level and/or the allowable adjusting margin together with the cost factor.

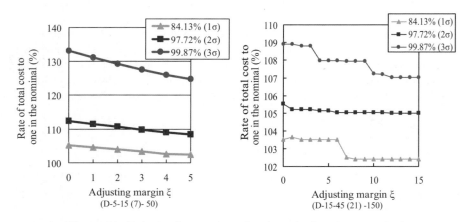

Fig. 2.33. Relation between total cost and adjusting margin ξ

2.6 Chapter Summary

In this chapter, we focused on a variety of single-objective optimization methods based on a metaheuristic approach.

These methods have emerged recently, and are nowadays filtering as practical optimization methods by virtue of the rapid progress of both computers and computer science. Roughly speaking, they are direct search methods aiming at a global optimum by utilizing a certain probabilistic drift. Their algorithms are characterized mainly by the ways in which to derive the tentative solution, how to evaluate it, and how to update it. They can even cope readily with the combinatorial optimization. Due to these favorable properties, these methods are being widely applied to some difficult problems encountered in manufacturing optimization.

To solve various complicated and large-scale problems in a numerically effective manner, we presented a hybrid approach that enables us to inherit the conventional outcomes and fuse them together with the recent outcomes straightforwardly. Types of hybrid approaches were classified, and an illustrative formulation was presented in terms of the combination of traditional mathematical programming and metaheuristic optimization in a hierarchical manner.

Then, three applications to manufacturing optimization were demonstrated to show how effectively each optimization method can solve each topic.

The first topic took a logistic problem associated with supply chain management that is closely related to the network design of hub systems such as transportation, telecommunication, *etc.* To deal with such large-scale and complex problems practically, a hybrid method was developed in a hierarchical manner. Through decomposing the problem into appropriate sub-problems, tabu search and the graph algorithm as a LP solver of the special class were applied to the resulting problems.

To increase the efficiency of the mixed-model assembly line for the small-lot-multi-kinds production, it is essential to prevent line stoppages incurred due to unexpected inconsistencies. The second topic concerned an injection sequencing problem under uncertainty associated with defective products. After formulating the problem, simulated annealing (SA) was employed to solve the resulting problem in a numerically effective manner.

Effective scheduling is one of the most important activities in intelligent manufacturing. However, little research has taken into account the role of human operators and cooperation between operators and resources. The third topic concerned production scheduling involving multi-skilled human operators manipulating multiple types of resources such as machine tools, robots and so on. A scheduling problem associated with human tasks was formulated and solved by an empirical optimization method known as the dispatching rule.

In the mathematical model employed for manufacturing optimization, there exist more or less uncertain factors that are impossible to forecast beforehand. In the last section, as a new interest related to the recent development of metaheuristic optimization methods, the application of GA to derive an insensitive solution against uncertain parameters was introduced. By virtue of its generic nature as a population-based algorithm, a high potential ability of coping with the uncertainty was examined through numerical experiments.

Then, focusing on the logistic systems associated with supply chain management, the hybrid tabu search was used again to solve the problem under uncertain customer demand. The idea from flexibility analysis was applied by classifying the decision variables as to whether they are soft (control) or hard (design). The results obtained there revealed that the approach is very promising for making a flexible logistic decision under uncertainties from comprehensive points of view.

References

1. Glover F W, Kochenberger GA (2003) Handbook of metaheuristics- variable neighborhood search (international series in operations research and management science 57). Springer, Netherlands
2. Ribeiro CC, Hansen P (eds.) (2002) Essays and surveys in metaheuristics. Kluwer, Norwell
3. Chambers LD (ed.) (1999) Practical handbook of genetic algorithms: complex coding systems. CRC Press, Boca Raton
4. Davis L (1991) Handbook of genetic algorithms. Van Nostrand Reinhold, New York
5. Goldberg DE (1989) Genetic algorithms in search, optimization and machine learning. Kluwer, Boston
6. Holland JH (1975) Adaptation in natural and artificial systems. University of Michigan Press, Ann Arbor
7. Kirkpatrick S, Gelatt CD, Vecchi MP (1983) Optimization by simulated annealing. Science, 220:671–680
8. Cerny V (1985) A thermodynamical approach to the traveling salesman problem: an efficient simulation algorithm. Journal of Optimization Theory and Applications, 45:41–51
9. Glover F (1989) Tabu search: Part I. ORSA Journal on Computing, 1:190–206
10. Glover F (1990) Tabu search: Part II. ORSA Journal on Computing, 2:4–32
11. Storn R, Price K (1997) Differential evolution–a simple and efficient heuristic for global optimization over continuous spaces. Journal of Global Optimization, 11:341–359
12. Kennedy J, Eberhart R (1995) Particle swarm optimization. Proc. IEEE International Conference on Neural Networks, pp. 1942–1948
13. Reynolds CW (1987) Flocks, herds, and schools: a distributed behavioral model, in computer graphics. Proc. SIGGRAPH '87, vol. 4, pp. 25–34
14. Dorigo M, Maniezzo V, Colorni A (1996) Ant system: optimization by a colony of cooperating agents. IEEE Transactions on Systems, Man, and Cybernetics-Part B, 26:29–41
15. Dorigo M, Stutzle T (2004) Ant colony optimization. MIT Press, Cambridge

16. Moscato P (1989) On evolution, search, optimization, genetic algorithms and martial arts: towards memetic algorithms. Caltech Concurrent Computation Program, C3P Report 826

17. Laguna M, Marti R (2003) Scatter search: methodology and implementation in C (Operations Research/Computer Science Interfaces Series 24). Kluwer, Norwell

18. Shimizu Y, Tachinami Y (2002) Parallel computing for solving mixed-integer programs through a hybrid genetic algorithm. Kagaku Kogaku Ronbunshu, 28:268–272 (in Japanese)

19. Karimi IA, Srinivasan R, Han PL (2002) Unlock supply chain improvements through effective logistic. Chemical Engineering Progress, 98:32–38

20. Knolmayer G, Mertens P, Zeier A (2002) Supply chain management based on SAP systems: order management in manufacturing companies. Springer, New York

21. Stadtler H, Kilger C (2002) Supply chain management and advanced planning: concepts, models, software, and case studies (2nd ed.). Springer, New York

22. Campbell JF (1994) A survey of network hub location. Studies in Locational Analysis, 6:31–49

23. Drezner Z, Hamacher HW (2002) Facility Location: applications and theory. Springer, New York

24. Ebery J, Krishnamoorth M, Ernst A, Boland N (2000) The capacitated multiple allocation hub location problem: formulations and algorithms. European Journal of Operational Research, 120:614–631

25. O'Kelly M E, Miller H J (1994) The hub network design problem. J. Transport Geography, 21:31–40

26. Lee H, Shi Y, Nazem SM, Kang SY, Park TH, Sohn MH (2001) Multicriteria hub decision making for rural area telecommunication networks. European Journal of Operational Research, 133:483–495

27. Wada T, Shimizu Y (2006) A hybrid metaheuristic approach for optimal design of total supply chain network. Transaction of ISCIE 19, 2:69–77 (in Japanese), see also Wada T, Shimizu Y, Yoo J-K (2005) Entire supply chain optimization in terms of hybrid in approach. Proc. 15th ESCAPE, Barcelona, Spain, pp. 591–1596

28. Okamura K ,Yamashida H (1979) A heuristic algorithm for the assembly line model-mix sequencing problem to minimize the risk of stopping the conveyor. International Journal of Production Research, 17:233–247

29. Yano C A, Rachamadugu R (1991) Sequencing to minimize work overload in assembly lines with product options. Management Science, 37:572–586

30. Yoo J-K, Moriyama T, Shimizu Y (2005) A sequencing problem in mixed-model assembly line including a painting line. Proc. ICCAS2005, Gyeonggi-Do, Korea, pp. 1118–1122

31. Pinedo M (2002) Scheduling: theory, algorithms, and systems (2nd ed.). Prentice Hall, Upper Saddle River

32. Blazewicz J, Ecker KH, Pesch E, Schmidt G, Weglarz J (2001) Scheduling computer and manufacturing processes (2nd ed.). Springer, Berlin

33. Muth JF, Thompson GL (1963) Industrial scheduling. Prentice Hall, Englewood Cliffs

34. Brucker P (2001) Scheduling algorithms. Springer, New York

35. Calrier J (1982) The one-machine sequencing problem. European Journal of Operation Research, 11:42–47

36. Iwata K et al. (1980) Jobshop scheduling with operators and proxy machines. Transactions of JSME, 417:709–718
37. Hino R, Kobayashi Y, Yoo J-K, Shimizu Y (2004) Generalization of scheduling problem associated with cooperation among human operators and machine. Proc. Japan–USA Symposium on Flexible Automation, Denver
38. Garcia-Flores R, Wang XZ, Goltz GE (2000) Agent-based information flow process industries' supply chain modeling. Computers & Chemical Engineering, 24:1135–1141
39. Gupta A, Maranas CD, McDonald CM (2000) Mid-term supply chain planning under demand uncertainty: customer demand satisfaction and inventory management. Computers & Chemical Engineering, 24:2613–2621
40. Zhou Z, Cheng S, Hua B (2000) Supply chain optimization of continuous process industries with sustainability considerations. Computers & Chemical Engineering, 24:1151–1158
41. Wada T, Yamazaki Y, Shimizu Y (2007) Logistic optimization using hybrid metaheuristic approach–consideration on multi-commodity and volume discount. Transactions of JSME, 73:919–926 (in Japanese), *see also* Wada T, Yamazaki Y, Shimizu Y (2007) Logistic optimization using hybrid metaheuristic approach under very realistic conditions. Proc. 17th ESCAPE, Bucharest, Romania, pp. 733–738
42. Hassin R (1983) The minimum cost flow problem: a unifying approach to existing algorithms and a new tree search algorithm. Mathematical Programming, 25:228–239
43. Shimizu Y, Wada T (2004) Hybrid tabu search approach for hierarchical logistics optimization. Transactions of ISCIE 17, 6:241–248 (in Japanese), *see also* Logistic optimization for site location and route selection under capacity constraints using hybrid tabu search. Proc. 8th International Symposium on PSE, pp. 612–617
44. Goldberg: AV (1997) An efficient implementation of a scaling minimum-cost flow algorithm. Algorithms, 22:1–29
45. http://www.ilog.co.jp
46. Miltenburg J (1989) Level schedules for mixed-model assemble lines in just-in-time production systems. Management Science, 35:192–207
47. Duplaga E A, Bragg DJ (1998) Mixed-model assembly line sequencing heuristics for smoothing component parts usage. International Journal of Production Research, 36:2209–2224
48. Monden Y (1991) Toyota production system: an integrated approach to Just-In-Time. Chapman & Hall, London
49. Koren Y, Heisel U, Jovane F, Moriwaki T, Pritshow G, Ulsoy G, Van BH (1999) Reconfigurable manufacturing systems. Annals of the CIRP, 48:527–540
50. Ruszczynski A, Shapiro A (eds.) (2003) Stochastic programming. Elsevier, London
51. Branke J (2002) Evolutionary optimization in dynamic environments. Kluwer, Norwell
52. Fitzpatrick JM, Grefenstette JJ (1988) Genetic algorithms in noisy environments. Machine Learning, 3:101–120
53. Hughes EJ (2001) Evolutionary multi-objective ranking with uncertainty and noise. In: Zitzler E et al.(eds.) EMO 2001. Springer, Berlin, pp. 329–343

54. Sano Y, Kita H (2002) Optimization of noisy fitness functions by means of genetic algorithms using history of search. Transactions of IEE Japan, 122-C, 6:1001–1008 (in Japanese)
55. Tamaki H, Arai T, Abe S (1999) A genetic algorithm approach to optimization problems with uncertainties. Transactions of ISCIE, 12:297–303 (in Japanese)
56. Adachi M, Yamamoto K, Shimizu Y (2003) A genetic algorithm for deriving insensitive solution against uncertain parameters. Proc. 46-th JAAC Conference, FA2-04-3, pp. 736–739 (in Japanese)
57. Jung JY, Blau G, Pekny JF, Reklaitis GV, Eversdyk D (2004) A simulation based optimization approach to supply chain management under demand uncertainty. Computers & Chemical Engineering, 28:2087–2106
58. Guillen G, Mele FD, Bagajewicz MJ, Espuna A, Puigjaner L (2005) Multiobjective supply chain design under uncertainty. Chemical Engineering Science, 60:1535–1553
59. Shimizu Y, Matsuda S, Wada T (2006) A flexible design for logistic network under uncertain demands through hybrid meta-heuristic strategy. Transactions of ISCIE, 19:342-349 (in Japanese), see also Flexible design of logistic network against uncertain demands through hybrid meta-heuristic method. Proc. 16th ESCAPE, Garmisch Partenkirchen, Germany, pp. 2051–2056
60. Swaney RE, Grossmann IE (1985) An index for operational flexibility in chemical process design. Part 1: formulation and theory. AIChE Journal, 31:621–630
61. Shimizu Y, Takamatsu T (1987) A design method for process systems with flexibility consideration. Kagaku Kogaku Ronbunshu, 13:574–580 (in Japanese)
62. Shimizu Y (1989) Application of flexibility analysis for compromise solution in large scale linear systems. Journal of Chemical Engineering of Japan, 22:189–194

3

Multi-objective Optimization Through Soft Computing Approaches

3.1 Introduction

Recently, agile and flexible manufacturing has been required to deal with diversified customer demands and global competition. The multi-objective optimization has been gaining interest as a decision aid sutable for those challenges. Accordingly, its importance might be intensified especially for real world problems in many fields. In this section, new methods for a multi-objective optimization problem (MOP)[1] will be presented associated with the metaheuristic methods and the soft computing techniques.

Generally, we can describe the MOP as a triplet like (x, f, x), similar to the usual single-objective optimization. However, it should be noticed that the objective function in this case is not a scalar but a vector. Consequently, the MOP is written, in general, by

$$[Problem] \quad \min \quad f(x) = \{f_1(x), f_2(x), \ldots, f_N(x)\}$$
$$\text{subject to} \quad x \in X,$$

where x denotes an n-dimensional decision variable vector, X a feasible region defined by a set of constraints, and f an N-dimensional objective function vector, some elements of which conflict and are incommensurable with each other.

The conflicts occur when if one tries to improve a certain objective function, at least one of the other objective functions deteriorates. As a typical example, if one weighs on the economy, the environment will deteriorate, and *vice versa*. On the other hand, the term incommensurable means that the objective functions lack a common scale to evaluate them under the same standard, and hence it is impossible to incorporate all objective functions into a single objective function. For example, environmental impact cannot

---

[1] A brief of review of the conventional methods is given in Appendix C.

be measured in terms of money, but money is usually used to account economic affairs.

To grasp the entire idea, let us illustrate the feature of MOP schematically. Figure 3.1 describes the contours of two objective functions f_1 and f_2 in a two-dimensional decision variable space. There, it should be noted that it is impossible to reach the minimum points of the two objective functions p and q simultaneously.

Here, let us make a comparison between three solutions, A, B and C. It is apparent that A and B are superior to C because $f_1(A) < f_1(C)$, and $f_2(A) = f_2(C)$, and $f_1(B) = f_1(C)$, and $f_2(B) < f_2(C)$. Thus we can rank the solutions from these comparisons. However, it is not true for the comparison between A and B. We cannot rank these as just the magnitudes of the objective values because $f_1(A) < f_1(B)$, and $f_2(A) > f_2(B)$. Likewise, a comparison between any solutions on the curve, $\overline{p - q}$, which is a trajectory of the tangent of both contour curves is impossible. These solutions are known as Pareto optimal solutions. Such a Pareto optimal solution (POS) becomes a rational basis for MOP since any other solutions are inferior to every POS. It should be also recalled, however, that there exist infinite POSs that are impossible to rank. Hence the final decision is left unsolved.

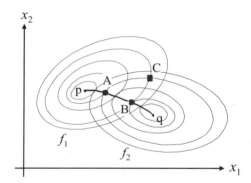

Fig. 3.1. Pareto optimal solution set in decision space, $\overline{p - q}$

To understand intuitively the POS as a key issue of MOP, it is depicted again in Figure 3.2 in the objective function space when $N = 2$. From this, we also know that there exist no solutions that can completely outperform any solution on the POS set (also called Pareto front) . For any solution belonging to the POS set, if we try to improve one objective, the rest of the objectives are urged to degrade as illustrated in the figure.

It is also apparent that it never provides a unique or final solution for the problem under consideration. For the final decision under multi-objectives, therefore, we have to decide a particular one among an infinite number of POSs. For this purpose, it is necessary to reveal a certain value function of

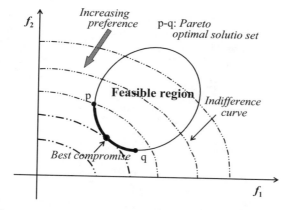

Fig. 3.2. Idea of a solution procedure in objective space

decision maker (DM) either explicitly or implicitly. This means that the final solution will be derived through the tradeoff analysis among the conflicting objectives by the DM. In other words, the solution process needs a certain subjective judgment to reflect the DM's preference in addition to the mathematical procedures. This is quite different from the usual or single-objective optimization problem (SOP) that will be completed only by mathematical procedures.

3.2 Multi-objective Metaheuristic Methods

As a suitable method associated with MOP, the extension of evolutionary algorithms (EA) has caused great interest. Strictly speaking, these methods are viewed as a multi-objective analysis that tries to reveal a certain feature of tradeoff among the conflicting objectives instead of aiming at obtaining a unique preferentially optimal solution. Such multi-objective evolutionary algorithm (MOEA) [1, 2, 3, 4] is an extension of EA in which the following two aspects are considered:

- How to select individuals belonging to the POS set.
- How to maintain diversity so that elements of POS set are derived not only as many as but also as varied as possible.

By considering multiple possible solutions simultaneously in search (population-based approach), MOEA can favorably generate a POS set in a single run of the algorithm. In addition, MOEA is less insensitive to the shape or continuity of the Pareto front (*e.g.*, they can deal with discontinuous and concave Pareto fronts without paying special attention). These are the special advantages[2] over the conventional mathematical programming techniques mentioned in Appendix C when dealing with the real world applications.

Below, only representative methods of MOGA will be outlined according to the following classification [5]:

• Aggregating function approach
• Population-oriented approach
• Pareto-based approach

3.2.1 Aggregating Function Approaches

The most straightforward approach of MOP is obviously to combine the multiple objective functions into a scalar one (a so-called aggregating function), and solve the resulting SOP using an appropriate method. Problem 3.1 with the linearly weighted sum objective function is one of the simplest dealing with this case,

$$[Problem] \quad \min \sum_{i=1}^{N} w_i f_i(x), \tag{3.1}$$

where $w_i \geq 0$ is a weight representing the relative importance among the N objectives, and is usually normalized such that $\sum_{i=1}^{N} w_i = 1$. Since EA needs scalar fitness information to work, a plain idea is to apply the above aggregating function value as a fitness. Though this approach is very simple and easy to implement, it has the disadvantage of missing concave portions of the Pareto front. Another difficulty is the determination of the appropriate weights to derive a global Pareto front when we do not have enough information about the problem *a priori*. These difficulties grow rapidly as the number of objective functions increases.

Goal programming, goal attainment, and the ϵ-constraint method are also available for the same purpose.

3.2.2 Population-oriented Approaches

To overcome the drawbacks of the aggregating methods, approaches in this class attempt to use the population-based effect of EA for maintaining the diversity of the search. They are known as the lexicographic ordering method [6], the method using gender to identify objectives [7] and randomly generated weight and elitism [8], the weighted min-max approach [9], non generational GA [10], *etc.*

The vector evaluated genetic algorithm (VEGA) proposed by Schaffer [11] is a classical method of this type. VEGA is a simple extension of the single-objective genetic algorithm with a modified selection mechanism. For a problem with N objectives, N sub-populations of size N_p/N each are generated

---

[2] Nevertheless, a comparison involving the computational load has been never discussed anywhere.

from a total population size of N_p. An individual in the sub-population, say k, is assigned a fitness based only on the k-th objective function. Using this value, the selection is performed per each sub-population. Since every member in the sub-population is selected based on the fitness of the particular objective function, its preference is consequently emphasized corresponding to the respective objective function. To generate a new population, genetic operations like crossover and mutation are applied after the sub-populations are merged together and shuffled to mix up. This procedure is illustrated in Figure 3.3.

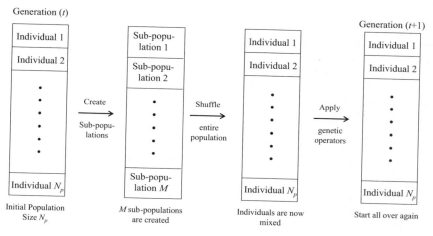

Fig. 3.3. Solution process of VEGA

Though this approach is easy enough to implement, some problems remain unsolved. Since the concept of Pareto optimality is not directly incorporated into the selection mechanism, the problem known as "speciation" arises. That is, let us suppose that the solution has a good compromise solution for all objectives ("middling" performance in all objectives), but it is not the best in any of them. Under this selection scheme, such a solution will hardly survive and be discarded nevertheless it could be very promising as a compromise solution.

Moreover, since merging and shuffling all sub-populations corresponds to averaging the fitness over the objective, the resulting fitness is substantially equivalent to a linear combination of the objectives. Hence, in the case of the concave Pareto front, we cannot attain the points on the concave portion by this method. Though it is possible to provide some heuristics to resolve these problems[3], the generic disadvantage associated with the selection mechanism remains.

3.2.3 Pareto-based Approaches

Under this category, we can incorporate the concept of Pareto optimality in the selection mechanism. Though various methods have been proposed in the last several years, only the representatives will be introduced below.

A. Non-dominated Sorting and the Multi-objective Genetic Algorithm (MOGA)

Methods in this class use a selection mechanism that favors solutions assigned high rank. Such ranking is performed based on non-dominance that aims at moving the population fast toward Pareto front. Once the ranking is performed, it is transformed into the fitness using an appropriate mapping function. All solutions with the same rank in the population are assigned the same fitness so that they all have the same probability of being selected.

Goldberg's ranking method [12, 13] is to find a set of solutions that are non-dominated by the rest of the population. Then, the solutions thus found are assigned the highest rank, say "1", and eliminated from further sorting. From the remaining populations, another set of solutions are determined and are assigned the next highest rank, say "2". This process continues until the population is suitably ranked. (see Figure 3.4). As is easily supposed, the performance of this algorithm will degrade rapidly as the increase in population size and the number of objectives. Goldberg also suggested the use of a niching technique [12] in terms of the sharing function so that the solutions cover the entire Pareto front.

In the case of ranking by Fonseca and Fleming [14], each solution is ranked based on the standard of how many other solutions will dominate it. When an individual x_i is dominated by $p_i(t)$ individuals in the current generation t, its rank is given by Equation 3.2.

$$\text{rank}(x_i, t) = 1 + p_i(t). \tag{3.2}$$

MOGA also uses a niche method to diversify the population. Though it can reduce the demerits of Goldberg's method and is relatively easy to implement, its performance is highly dependent on an appropriate selection of sharing parameter σ_{share} that can adjust the niche. This property is common to all other Pareto ranking techniques.

---

[3] For example, add a few linearly weighted sum objectives with different weighting coefficients, *i.e.*, $f_{N+j}(x) = \sum_{i=1}^{N} w_i^j f_i(x), (j = 1, 2, \ldots)$ to the original objective functions.

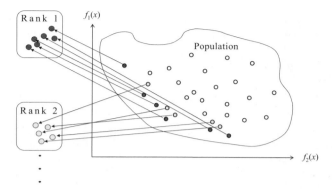

Fig. 3.4. Solution process of Goldberg's ranking method

B. The Non-dominated Sorting Genetic Algorithm (NSGA)

Before the selection is performed, NSGA [15] ranks population N_p into mutually exclusive non-dominated sets P_i on the basis of a non-domination concept,

$$N_p = \bigcup_{i=1}^{K} P_i,$$

where K is the number of non-dominated sets. This will classify the population into several layers of fronts as depicted in Figure 3.5.

Then the fitness assignment procedure takes place from the most preferable front ("1") to the least ("K") in turn. First, a fitness equal to the population size N_p is given to all solutions on front "1" to provide an equal probability of selection, *i.e.*, $F_i = N_p, (\forall i \in$ Front 1). To maintain the diversity among the solutions in the front, the assigned fitness above is degraded in terms of

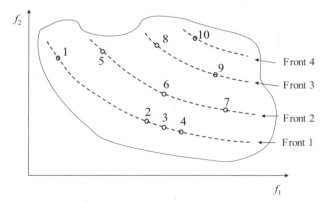

Fig. 3.5. Idea of non-dominated sorting

the number of neighboring solutions, or sharing concept. For this purpose, the normalized Euclidean distance from another solution in the same front is calculated in the decision variable space. Then, by applying this value to the sharing function and obtaining niche count nc_i, the shared fitness is evaluated as $\tilde{F}_i = F_i/nc_i$. Next moving to the second front, we assign the fitness of all solutions on this front at the value slightly smaller than the minimum shared fitness at the first front, i.e., $\min_{i \in \text{Front } 1} \tilde{F}_i - \epsilon$, and obtain the shared fitness based on the same procedures mentioned above. This process is continued until all layers of the front are considered. Since the solutions in the preferable front have a greater fitness value than the less preferable ones, they are always likely to be reproduced compared with the rest of the individuals in the population.

C. Niched Pareto Genetic Algorithm (NPGA)

This method [16] employs a selection mechanism called Pareto domination tournament. First, a pair of solutions (i, j) are chosen at random in the population and they are individually compared with every member of a sub-population T_{ij} of size t_{dom} based on the non-domination concept. If i is non-dominated by the samples and j is not, the i becomes a winner, and vice versa (see also Figure 3.6). If there is a tie (both are either dominated or non-dominated), then the sharing strategy will decide the winner. (At the beginning, this step will be skipped, and i or j is chosen with equal probability, i.e., 0.5.) Based on the normalized Euclidian distance in the objective function space between i or j and $k \in Q$ (offspring population), the niche counts nc_i and nc_j are computed. If $nc_i \leq nc_j$, solution i becomes the winner, and vice versa. The above procedures are repeated again, and each winner becomes the next parents that will create a new pair of offspring through the genetic operators, i.e., crossover and mutation. This cycle will be continued to fill the population size of offspring by N_p. Since this approach applies the non-dominated sorting only to the limited sub-population and dynamically updated niching, it is very fast and produces good non-dominated solutions that can be kept for a large number of generations. Moreover, it is unnecessary to specify any particular fitness value to each solution. However, the good performance of this approach greatly depends on a good choice of value t_{dom} as well as the sharing factor or niche count.

D. The Elitist Non-dominated Sorting Genetic Algorithm (NSGA-II)

NSGA-II [17], a variant of NSGA, uses the idea of elitism that can avoid both deleting the superior solutions found previously and crowding to maintain the diversity of solutions. In this method, non-dominated sorting is carried out for all members of the parents $P(t)$ and offspring $Q(t)$ populations (hence, a total of $2M$ solutions are considered.). To create the parent population of size N_p at the next generation $P(t + 1)$, solutions on each front are filled in order of preference class by reaching the size of N_p. Generally, since it is impossible to fill all members in the last class, a crowding distance is used to decide the

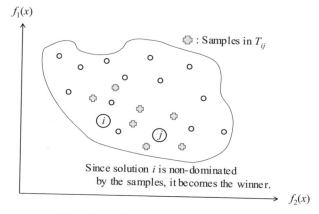

Fig. 3.6. Solution process of NPGA

members included in the population as depicted in Figure 3.7. The crowding distance is an estimate of the density of solutions neighboring a particular solution:

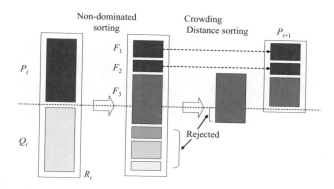

Fig. 3.7. Solution process of NSGA2

Then the offspring population $Q(t+1)$ is created from $P(t+1)$ by using the crowded tournament selection, crossover and mutation operators. Relying on the non-dominated rank and local crowding distance, solution i wins solution j if either of the following conditions is satisfied (the crowded tournament selection).

- Solution i belongs to a more preferable rank than solution j.
- When they are tied, the crowding distance of solution i is greater than that of j.

By virtue of these operators, NSGA-II is considerably faster than its prede-
cessor NSGA and gives very good results for many problems.

F. Miscellaneous

Besides the methods described above, a variety of methods have been proposed. For example, the vector optimized evolution strategy (VOES) [18],
and the predator-prey evolution strategy [19] are non-elitist algorithms in the
Pareto-based category. On the other hand, the distance-based Pareto genetic
algorithm (DPGA) [20], the strength Pareto evolutionary algorithm (SPEA)
[21], the multi-objective messy genetic algorithm (MOMGA) [22], the Pareto
archived evolution strategy (PAES) [23], the multi-objective micro-genetic
algorithm (MμGA) [5], and the multi-objective program (GENMOP) [24] belong to elitist algorithms.

A comparison of multi-objective evolutionary algorithms was made, and
revealed that elitism plays an important role in improving evolutionary multi-
objective search [25]. Moreover, regarding other meta-approaches besides GA,
multi-objective simulated annealing [26] is known, and the concept of non-
dominated sorting and a niche strategy are applied in tabu search [27]. Also,
extensions of DE are proposed in recent studies [28, 29]. A multi-objective
scatter search is applied to solve a mixed-model assembly line sequencing
problem [30].

Unfortunately, all these algorithms give only a set of solutions though we
are willing to have at most several candidate solutions in real world applications. This is because MOEA is not of concern about any preference information imbedded by the DM, and highlights the diversity of solutions over the
entire Pareto front as a technique for multi-objective analysis. However, even
in multi-objective analysis, we should address the interest of the DM's preference more elaborately. Let us consider this problem by taking the following
ϵ-constraint method as an example:.

$$[Problem] \quad \min \ f_p(x) \quad \text{subject to} \quad f_i(x) \leq f_i^* + \varepsilon_i, \quad (i = 1, 2, \ldots, N, i \neq p).$$

If a value function of that the DM conceived implicitly is described by $V(f(x))$,
the multi-objective analysis must be concentrated within the particular extent
that the DM prefers. According to this intention, the above problem should
be re-described as

$$[Problem] \quad \min \ f_p(x) \quad \text{subject to} \quad \begin{cases} f_i(x) \leq f_i^* + \varepsilon_i \ (i = 1, \ldots, N, i \neq p) \\ \partial V/\partial f_i \leq 0 \ (i = 1, \ldots, N, i \neq p). \end{cases}$$

In terms of this idea, a discussion of diversification is meaningful over the
entire front in the case of (a) in Figure 3.8, because the preference will increase
everywhere on the front if we reduce either objective functions. In the other
case (b) under a different value system, it is enough to emphasize the diversity
only in the limited extent of the front crossing with the painted triangle in

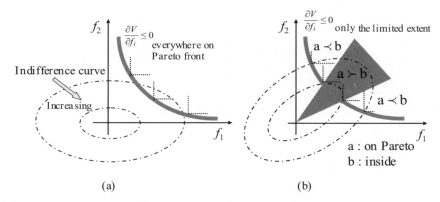

Fig. 3.8. Two cases of meaningful Pareto front: (a) over the entire front, (b) in the limited front

the figure. This is because we can obtain a more preferable solution by leaving from the front outside of this region.

How to deal with problems with more than three objectives may be another difficulty remaining unresolved for MOEA. This is easily supposed from the fact that the simple schematic representation of the Pareto front is impossible for $N > 3$.

3.3 Multi-objective Optimization in Terms of Soft Computing

As mentioned in Chap. 1, soft computing (SC) is a collection of computational techniques in computer science, artificial intelligence, and machine learning. The major areas of SC are composed of neural networks, fuzzy systems and evolutionary computation. SC has more tolerance regarding imprecision, uncertainty, partial truth, and approximation; and makes a larger point on the inductive reasoning than conventional computing. Moreover, new hybrid approaches are expected to be invented by a particularly effective combination of SC. The multi-objective optimization method mentioned below presents a new type of approach that may facilitate significant computing technologies targeted at manufacturing systems.

Let us describe MOP in the general form again,

$$[Problem] \quad \min \ f(x) \ = \{f_1(x), f_2(x), \dots, f_N(x)\}$$
$$\text{subject to } x \in X. \tag{3.3}$$

As mentioned already, we need some information on the DM's preference to attain the preferentially optimal solution of MOP in addition to the mathematical procedures. To avoid a certain stiffness and shortcomings encountered in conventional methods, a few multi-objective optimization methods in terms of soft computing (MOSC) will be presented below. They are called multi-objective hybrid GA (MOHybGA [31]), the multi-objective optimization method with a value function modeled by a neural network [32, 33] (MOON^2) and MOON^{2R} [34, 35], MOON^2 of radial basis function. These methods can derive a unique solution that is the best compromise of DM. Due to this fact, they are expected to be powerful tools for flexible decision making for agile manufacturing.

3.3.1 Value Function Modeling Using Artificial Neural Networks

Since these methods belong to a prior articulation method of MOP, they needs to identify a value function of DM *a priori*. To deal with the non-linearity commonly embedded in the value function, the artificial neural network[4] is favorably available for such modeling. A back propagation (BP) network is used in MOHybGA and MOON^2, while MOON^{2R} employs a radial basis function (RBF) network [4]. The RBF network is more flexible and easier than the BP network regarding the training and dynamic adaptation against incremental operations due to the change of neural network structure. That enables us to model the value function more readily, depending on the unsteady decision environment often encountered in real world problems.

To train the neural network, data standing for the preference of DM should be gathered in an appropriate manner. These methods use pair-wise comparisons among the trial solutions that are composed of several reference solutions spread over the search area in the objective function space. It is natural to constrain the modeling space within the hull convex enclosed by the utopia and nadir solutions. For example, $f^{\text{utop}} = (f_1(x^{\text{utop}}), f_2(x^{\text{utop}}), \ldots, f_N(x^{\text{utop}}))^T$ and $f^{\text{nad}} = (f_1(x^{\text{nad}}), f_2(x^{\text{nad}}), \ldots, f_N(x^{\text{nad}}))^T$, where x^{utop} and x^{nad} are utopia and nadir solutions in decision variable space, respectively. Several methods to set up these reference solutions are known.

1. Ask the DM to reply his/her selections directly.
2. Set up them referring to the pay-off table[5].
3. Do this in combination with the above, *i.e.*, the utopia from the pay-off table, and the nadir from the response from the DM.

The rest of the trial solution f^s may be generated randomly so that they do not locate too closely to each other. For example, it is generated successively as follows:

$$f^s = f^{\text{utop}} + \text{rand}()(f^{\text{nad}} - f^{\text{utop}}), \qquad (3.4)$$
$$\| f^s - f^t \| \geq d, (t = 1, \ldots, k, t \neq s),$$

---

[4] The basis of the neural network named here is outlined in Appendix D.
[5] Refer to Appendix C for the construction of the pay-off matrix.

Table 3.1. Conversion table

Linguistic statement	a_{ij}
Equally	1
Moderately	3
Strongly	5
Very strongly	7
Extremely	9
Intermediate judgments 2,4,6,8	

where f^t denotes the solutions derived previously, rand ( ) a random number in [0,1], and d a threshold to keep distance between the adjacent trial solutions (refer to Figure 3.9).

Then, the DM is asked to reply which one he/she likes, and what the degree is between every pair of the trial solutions, say f^i and f^j. Such responses will take place by using the linguistic statements, and later transformed into the score a_{ij} as shown in Table 3.1, which is the same as AHP [29]. For example, when the answer is such that f^i is strongly preferable to f^j, a_{ij} becomes 5.

When the number of objectives is at most three, this is a rather easy way to extract the DM's preference. Especially, it should be noticed that this pair-wise comparison can be performed more adequately than the pair-wise comparison in AHP. That is, though we are alien to the comparison between the abstract attributes, *e.g.*, the importance between "swiftness" and "cost", we are used to the comparison between the candidates with concrete attribute values, *e.g.*, attractiveness between K-rail = {swiftness:2 hrs, cost: 4000 yen} and J-rail = {swiftness:1 hr, cost: 6000 yen} to buy a train ticket. In fact, this kind of pair-wise comparison is very often encountered in our daily life.

After doing such pair-wise comparisons over k trial solutions in turn, we can obtain a pair-wise comparison matrix (PWCM) as shown in Figure 3.10. Its (i, j) element a_{ij} represents a degree of preference of f^j compared with f^i stated using a certain score in Table 3.1. It is defined as the ratio of the

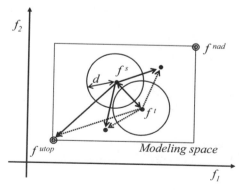

Fig. 3.9. Generation method of trial solutions (two-objective problem)

	f^1	f^2	f^3	• •	f^k
f^1	1	a_{12}	a_{13}	• •	a_{1k}
f^2		1	a_{23}	• •	a_{2k}
f^3			1	• •	⋮
⋮	$a_{ij} = 1/a_{ji}$			•.	⋮
f^k					1

Fig. 3.10. Pair-wise comparison matrix

relative degree of preference, but it does not necessarily mean f^i is a_{ij} times preferable to f^j. According to the same conditions as AHP, such that $a_{ii} = 1$ and $a_{ji} = 1/a_{ij}$, DM is required to reply $k(k-1)/2$ times in total. Under these conditions, it is also easy to examine the consistency of such pair-wise comparisons from the consistency index CI adopted in AHP,

$$CI = (\lambda_{\max} - k)/(k - 1), \tag{3.5}$$

where $\lambda_{\max}$ denotes the maximum eigenvalue of PWCM. It is empirically known if CI exceeds 0.1, there are undue responses involved in the matrix. In such a case, we need to revise certain scores to fix the inconsistency problem.

Generally speaking, it is almost impossible to give a mathematically definite form to the value function that is highly nonlinear. Since the preference information of DM is imbedded in the PWCM, it is relevant to derive a value function based on it. Under such understanding, a unstructured modeling technique using neural networks is known to be suitable for such modeling. PWCM provides a total of k^2 training data for the neural network. That is, all objective values of every pair, say f^i and f^j, $(\forall i, j \in \{1, 2, \ldots, k\})$ become $2N$ inputs, and the (i, j) element of PWCM a_{ij} an output of the neural network. Thus a trained neural network using these data can be viewed as an implicit function mapping $2N$-dimensional space to scalar space, $i.e.$, $V_{NN} : (f^i(x), f^j(x)) \in \mathrm{R}^{2N} \rightarrow a_{ij} \in \mathrm{R}^1$. Furthermore, let us notice the following relation:

$$V_{NN}(f^i, f^k) = a_{ik} \geq V_{NN}(f^j, f^k) = a_{jk}$$
$$\Leftrightarrow f^i \succeq f^j, \ (\forall i, j, k). \tag{3.6}$$

Then, we can rank the preference of any solutions by the output of the neural network, $a_{*\mathrm{R}}$. It is calculated by fixing one of the input vectors at an appropriate reference, say f^{R},

$$a_{*\mathrm{R}} = V_{NN}(f(x), f^{\mathrm{R}}).$$

In other words, trajectories with the same output value of a_{*R} are equivalent to the indifference curves or contours of the value function in the objective space. Such assertion is valid as long as the consistency of the pair-wise comparison is satisfied (*i.e.*, $CI < 0.1$). Numerical experiments using a few test problems reveal that a few typical value functions can be modeled correctly by a reasonable number of pair-wise comparisons [31].

Now, Problem 3.3 can be transformed into the following SOP:

$$[Problem] \quad \max \ V_{NN}(f(x), f^{\mathrm{R}}) \ \text{subject to} \ x \in X. \quad (3.7)$$

The following proposition supports the validity of the above formulation.

[Proposition] The optimal solution of Problem 3.8 is a Pareto optimal solution of Problem 3.3 if the value function is identified so as to satisfy the relation given by Equation 3.6.

(Proof) Let $\hat{f}_i^*, (i = 1, \ldots, N)$ be a value of each objective function for the optimal solution $\hat{x}^*$ of Problem 3.8, *i.e.*, $\hat{f}_i^* = f_i(\hat{x}^*)$.

Here, let us assume that $\hat{f}^*$ is not a Pareto optimal solution. Then there exists f^0 such that for $\exists j, \ f_j^0 < \hat{f}_j^* - \Delta f_j, (\Delta f_j > 0)$ and $f_i^0 \leq \hat{f}_i^*, (i = 1, \cdots, N, i \neq j)$. Since DM obviously prefers f^0 to $\hat{f}^*$, it holds that $V_{NN}(f^0, f^{\mathrm{R}}) > V_{NN}(\hat{f}^*, f^{\mathrm{R}})$. This contradicts that $\hat{f}^*$ is the optimal solution of Problem 3.8. Hence $\hat{f}^*$ must be a Pareto optimal solution.

Regarding the setting of reference point f^{R}, we can nominate some candidates such as utopia, nadir, a center of gravity between them, and the point where the total sum of distance from all trial points becomes minimum. Since there exist no definite theoretical backgrounds for such a selection, the following procedure similar to the successive linear approximation of function may be amenable to improving the quality of solution.

Step 1: Obtain a tentative solution by setting the reference point at the nadir point.

Step 2: Reset the reference to the foregoing tentative solution.

Step 3: Derive the updated solution.

Step 4: Repeat these procedures until the consecutive solutions coincide with each other with the admissible extent.

3.3.2 Hybrid GA for Solving MIP under Multi-objectives

This section describes an extension of the hybrid GA presented in Sect. 2.3 to solve MIP under multi-objectives (MOMIP) in terms of the foregoing modeling technique of the value function. The problem under consideration is given as follows:

$$[Problem] \quad \min_{x,z} \{f_1(x, z), f_2(x, z), \ldots, f_N(x, z)\},$$

$$\text{subject to} \quad \begin{cases} g_i(x, z) \geq 0 & (i = 1, \ldots, m_1) \\ h_i(x, z) = 0, & (i = m_1 + 1, \ldots, m) \\ x \geq 0, & \text{(real)} \\ z \geq 0, & \text{(integer)} \end{cases},$$

where x and z represent an n-dimensional real value vector and an M-dimensional integer value vector, respectively.

In addition to the multiple objectives, the existence of both integer and real variables should be notable in this problem. To derive the POS set of MOMIP, the following hierarchical formulation is possible:

$$[Problem] \quad \min_{z \geq 0:\text{integer}} f_p(x, z)$$

$$\text{subject to} \quad \min_{x \geq 0:\text{real}} f_p(x, z),$$

$$\text{subject to} \quad \begin{cases} f_i(x, z) \leq f_i^* + \epsilon_i & (i = 1, \ldots, N, i \neq p) \\ g_i(x, z) \geq 0 & (i = 1, \ldots, m_1) \\ h_i(x, z) = 0 & (i = m_1 + 1, \ldots, m) \end{cases},$$

where $f_p(\cdot)$ denotes a principal objective function, f_i^* the optimal value of the i-th objective, and ϵ_i its amount of degradation. In the above, the lower level problem refers to the usual ϵ-constraint problem, which derive a Pareto optimal solution even in the non-convex case. Moreover, to deal with this hierarchically formulated scheme, the hybrid approach below is known to be amenable. By solving this problem for a variety of ϵ_i, the POS set can be derived in a systematic way.

As is commonly known, the best compromise solution should be chosen from the POS set at the final step of MOP. For this purpose, an appropriate tradeoff analysis among the candidate solutions becomes necessary. Eventually, such a tradeoff analysis refers to a process to adjust the attained level of each objective value according to the DM's preference. In other words, in the above formulation, the best compromise solution is obtained by deciding the most preferable amounts of degradation of the objective value, *i.e.*, ϵ_i. To make such a decision, the following idea is suitable:

Step 1: Define an unconstrained optimization problem to search integer variables and quantized amounts of degradation by GA.

Step 2: Solve the constrained optimization problem regarding real variables by a certain mathematical programming (MP) while pegging the integer variables at the values decided at the upper level.

Step 3: Return to the upper level with the optimized real variables.

Step 4: Repeat the procedures until a certain stopping condition has been attained.

Such a scheme can bring about a good match between the solution methods and the properties of the problems, *i.e.*, GA with the unconstrained combinatorial optimization, and MP with the constrained continuous one. However, the usual application of GA accompanies much subjective judgment of the DM, which is actually impossible. To get rid of this inconvenience, the scheme formulated below is suitable for applying a hybrid method of GA and MP under multi-objectives. (see also Figure 3.11)

$$[Problem] \quad \max_{z, \epsilon_{-p}} \ V_{NN}(\epsilon_{-p}, f_p(x, z); f^R)$$

$$\text{subject to } \min_{x:\text{real}} f_p(x, z),$$

$$\text{subject to } \begin{cases} f_i(x, z) \leq f_i^* + \epsilon_i \ (i = 1, \ldots, N, i \neq p) \\ g_i(x, z) \geq 0 \ (i = 1, \ldots, m_1) \\ h_i(x, z) = 0 \ (i = m_1 + 1, \ldots, m) \end{cases},$$

where ϵ_{-p} means a vector composed of the ϵ-constrained amount of every element except for the p-th one, *i.e.*, $\epsilon_{-p} = (\epsilon_1, \ldots, \epsilon_{p-1}, \epsilon_{p+1}, \ldots, \epsilon_N)^T$, and V_{NN} a value function identified through the pair-wise comparison between two candidate solutions, *i.e.*, (ϵ_{-p}^i, f_p^i) and (ϵ_{-p}^j, f_p^j)[6]. The detail of the algorithm is described below on the basis of the simple GA [13].

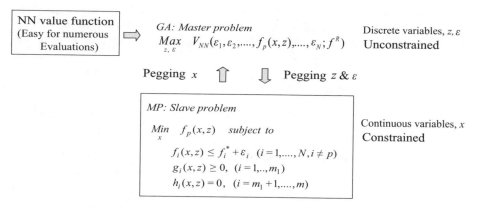

Fig. 3.11. Scheme of hybrid GA under multi-objectives

A. Chromosome Representation

Figure 3.12 shows a binary representation whose front half corresponds to the integer variables, and the rear half to the quantized amounts of degradation of ϵ-constraints. They are decoded, respectively, as follows:

---

[6] Considerations on the inactive ϵ-constraints in the lower level problem are discussed in the literature [38].

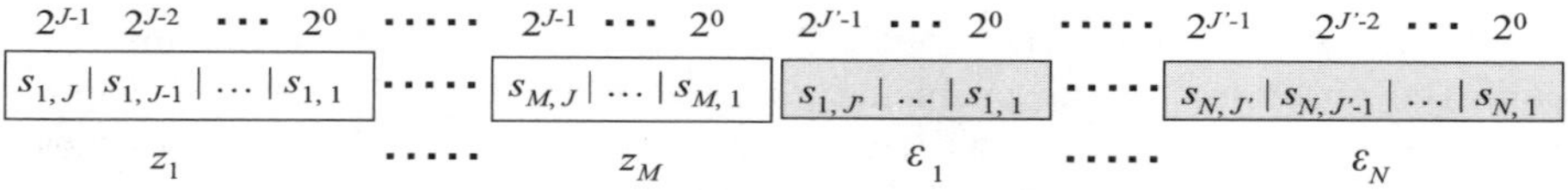

Fig. 3.12. Chromosome of hybrid GA

$$z_i = \sum_{j=1}^{J} 2^{j-1} s_{ij} \quad (i = 1, \ldots, M),$$

$$\epsilon_i = \sum_{j=1}^{J'} 2^{j-1} s_{ij} \delta \epsilon_i \quad (i = 1, \ldots, N, i \neq p),$$

where each s_{ij} denotes a 0-1 variable representing the binary type of allele, and $\delta \epsilon_i$ a grain of quantization[7]. Moreover, J and J' denote the number of bits necessary to prescribe the interval of variables regarding z_i and ϵ_i, respectively. Integer variables can be precisely expressed by such a binary coding. In contrast, the binary coding of real variables exhibits a tradeoff problem between the efficiency (chromosome length) and the accuracy (grain size). However, the present binary coding for real ϵ_i is a relevant representation since people usually have a certain resolution magnitude that can identify the preference difference between two solutions. Hence, its grain size $\delta \epsilon_i$ can be decided almost automatically. These facts support the adequateness of the coding in this hybrid approach.

B. Genetic Operators

Reproduction: The usual roulette wheel strategy is employed in the application [31].

Crossover: The usual one-point crossover per each part as shown in Figure 3.13 is simple and relevant (virtually two-points crossover).

Mutation: The usual binary bit entry flip (*i.e.*, 0/1 or 1/0) is simple and relevant.

Evaluation of fitness: The output of the value function modeled by using a neural network is transformed properly in terms of an appropriate scaling function to calculate the fitness value.

Moreover, relying on the nature of the population-based approach of GA, the above formulation is applicable to an ill-posed problem where the relevant objectives under consideration consist of both quantitative and qualitative

---

[7] If the variable on the interval $[0, 10]$ is described by the 4-bit length of the chromosome, this becomes $\delta \epsilon_i = (10 - 0)/(2^4 - 1)$.

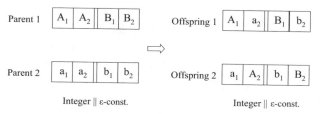

Fig. 3.13. Crossover of MOHybGA

objectives, *e.g.*, [39]. Since the direct evaluation or the metric evaluation is generally impossible for the qualitative objectives, it is rational to choose only tentatively several promising candidates from the quantitative evaluation, and leave the final decision to be based on the comprehensive evaluation by DM. Such an approach can be easily realized by computing the transformed fitness using a sharing function [13].

First, for the chromosome coded as shown in Figure 3.12, the Hamming distance between m and n, d_{mn} is computed by

$$d_{mn} = \sum_{i=1}^{M} \sum_{j=1}^{J} \mid s_{ij}(m) - s_{ij}(n) \mid + \sum_{\substack{i=1 \\ i \neq p}}^{N} \sum_{j=1}^{J'} \mid s_{ij}(m) - s_{ij}(n) \mid,$$

where $s_{ij}(\cdot)$ denotes the allele of the chromosome (binary code, *i.e.*, 0 or 1).

After normalizing d_{mn} by the length of chromosome as $\hat{d}_{mn} = d_{mn}/(JM + J'(N-1))$, the modified (shared) fitness $\hat{F}_m$ is derived from the original F_m as

$$\hat{F}_m = F_m / \sum_{n=1}^{N_p} \{1 - (\hat{d}_{mn})^a\} \quad (m = 1, \ldots, N_p),$$

where $a(> 0)$ is a scaling coefficient and N_p the population size.

Using the shared fitness, it is possible to generate various near-optimal solutions that locate around the optimal one while being somewhat distant from each other. These alternatives can have nearly the same fitness value evaluated only by the quantitative objective function, but they are expected to have a variety of bit patterns of the code due to the sharing operation. Hence, there might exist several solutions that are individually different from the qualitative evaluation. Consequently, a final decision is to be made by inspecting these alternatives carefully through adding evaluation from the qualitative objectives.

3.3.3 MOON^{2R} and MOON^2

A. Algorithm of MOON^{2R}

As shown already, the original MOP can be transformed into a SOP once the value function is modeled using a neural network. Hence it is applicable to a variety of optimization methods known previously. The difference between $MOON^2$ and $MOON^{2R}$ (together termed MOSC, hereinafter) is only the type of neural network employed for value function modeling, though the RBF network is more adaptive than the BP network. The following statements are developed on a case-by-case basis. Accordingly, the resulting SOP in $MOON^{2R}$ is rewritten as follows.

$$[Problem] \quad \max \quad V_{\mathrm{RBF}}(f(x), f^{\mathrm{R}}) \quad \text{subject to } x \in X. \tag{3.8}$$

When this approach is applied with the algorithm that requires gradients of the objective function such as nonlinear programs, they need to be obtained by numerical differentiation. The derivative of the value function with respect to the decision variable is calculated from the following chain rule:

$$\left(\frac{\partial V_{\mathrm{RBF}}(f(x), f^R)}{\partial x} \right) = \left(\frac{\partial V_{\mathrm{RBF}}(f(x), f^R)}{\partial f(x)} \right) \left(\frac{\partial f(x)}{\partial x} \right). \tag{3.9}$$

With the analytic form of the second part in the right-hand side of Equation 3.9 and the following numerical differentiation, the calculation of the derivative can be completed,

$$\left(\frac{\partial V_{\mathrm{RBF}}(f(x), f^R)}{\partial f_i(x)} \right) = (V_{\mathrm{RBF}}(f_1(x), \ldots, f_i(x) + \Delta f_i, \ldots, f_N(x), f^R)$$
$$- V_{\mathrm{RBF}}(f_1(x), \ldots, f_i(x) - \Delta f_i, \ldots, f_N(x), f^R))/2\Delta f_i, \ (i = 1, \ldots, N). \tag{3.10}$$

Since most nonlinear programming software support numerical differentiation, the algorithm is achieved without any special problems.

Moreover, any candidate solutions can be evaluated readily under the multi-objectives through V_{RBF} once x is given. Hence we can engage in MOP by just using an appropriate method among a variety of conventional methods for SOP. Not only direct methods but also metaheuristic methods like GA, SA, tabu search, *etc.* are readily applicable. In contrast, any interactive methods of MOP are almost impossible to apply because they require too many interactions making DM disgust and slipshod during the search.

Figure 3.14 shows a flowchart the procedure of which is outlined as follows:

Step 1: Generate several trial solutions in the objective function space.
Step 2: Extract DM's preference through pair-wise comparison between every pair of the trial solutions.
Step 3: Train a neural network based on the preference information obtained from the above responses. This derives a value function V_{NN} or V_{RBF}.
Step 4: Apply an appropriate optimization method to solve the resulting SOP, Problem 3.8.

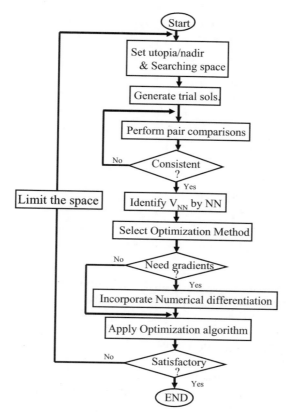

Fig. 3.14. Flow chart of the proposed solution procedure

Step 5: If DM is unsatisfied with the result obtained above, limit the search space around there, and repeat the same procedure until the result is satisfactory.

In this approach, since the modeling process of the value function is separated from the search process, the DM can carry out tradeoff analyses at his/her own pace without worrying about the hurried and/or idle responses often experienced with the interactive methods. In addition, since the required responses are simple and relative, the DM's load in such an interaction is very small. These are special advantages of this approach.

However, since the data used for identifying the value function is obtained from human judgment on preference, it is subjective and not rigid in a mathematical sense. In spite of this, MOSC can solve MOP under a variety of preferences effectively as well as practically. This is because MOSC is considered to be robust against the element value of the PWCM just like AHP. In addition, the optimality can be achieved on an ordinal basis rather than a cardinal one, or it is not so sensitive with respect to the shape of the function, as illustrated in Figure 3.15.

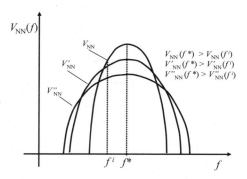

Fig. 3.15. Insensitivity against the entire shape of the value function

However, inadequate modeling of the value function is likely to cause an unsatisfactory result at Step 5 in the above procedure. Moreover, the complicated nonlinearity of the value function and changes of decision environment can sometimes alter the preference of the DM. Such a situation requires us to modify the value function adaptively. Regarding this problem, the RBF network also has a nice property. Its retraining easily takes place through incremental operations against both increase and decrease in the training data and the basis from the foregoing one as shown in Appendix D.

B. Application in an Ill-posed Decision Environment

Being closely related to the nature of people, there are many cases where the subjective judgment such as the pair-wise comparison may involve various confusions due to misunderstandings and/or unstable decision circumstance at work. The more pair-wise comparisons DM needs to make, the more likely it is that a lack of concentration and misjudgments will be induced in terms of simple repetition of the responses. To facilitate a wide application of MOSC, therefore, it is necessary to cope with such problems adequately and practically as well. Classifying such improper judgments into the following three cases, let us consider the methods to find out the irrelevant responses in the pair-wise comparisons, and revise them properly [40].

1. The case where the transitive relation on preference will not hold.
2. The case where the pair-wise comparison may involve over preferred and/or underpreferred judgments.
3. The case where some pair-wise comparisons are missing.

Case 1 occurs when preferences among three trials f^i, f^j, f^k result in such relations that the DM prefers f^i to f^j, f^j to f^k, and f^k to f^i. On the other hand, Case 2 corresponds to the situation where the judgment on preference differs greatly from the true one due to an overestimate or an underestimate. When $f^i \prec\prec f^j$, the response such as $f^i \prec f^j$ is an example of the overpreferred judgment of f^i to f^j, or equivalently to say, the underpreferred one of f^j to f^i. Here, notations $\prec$ and $\prec\prec$ mean the relation that is preferable and very preferable, respectively. By calculating the consistency index CI defined by Equation 3.5, we can find the occurrence of these defects since such responses will degrade CI considerably. If CI exceeds the threshold value (usually, 0.1), the following procedures are available to fix the problems.

For the first case, we can apply the level partition of ISM method (interpretive structural modeling [2]) after transforming PWCM into the quasi-binary matrix as shown in Appendix E. From the result, we can detect the inconsistent pairs, and ask the DM to evaluate them again.

Meanwhile, we cope with Case 2 as follows.

Step 1: First compute the weights $w_i(i = 1, \ldots, k)$ representing the relative importance among the trial solutions from PWCM ($\{a_{ij}\}$) using the same procedure as AHP.

Step 2: Obtain the completely consistent PWCM $\{a_{ij}^*\}$ from the weights derived in Step 1, i.e., $a_{ij}^* = w_i/w_j$.

Step 3: Compare every element of (the inconsistent) PWCM with each of the completely consistent matrix, and find some elements that are far apart from with each other, i.e., the m-biggest $|a_{ij}^* - a_{ij}|/a_{ij}^*$, $(\forall\ i, j)$.

Step 4: Fix the problem in either of the following two ways.

1. Ask the DM to reevaluate the identified undue pair-wise comparisons.
2. Replace the worse elements, say a_{ij} with the default value, i.e., $\min\{a_{ij}^*, 9\}$ if $a_{ij}^* \geq 1$, or $\max\{a_{ij}^*, 1/9\}$ if $a_{ij}^* < 1$.

Moreover, Case 3 occurs when the DM cannot decide his/her attitude immediately or suspend it due to certain tedious correspondences associated with the repeated comparison. Accordingly, some missing elements are involved in the PWCM. We can cope with this problem by applying the method known as Harker's method [42]. It relies on the fact that the weight can be calculated only from a part of PWCM if it is completely consistent. Hence, after calculating the weight even from the incomplete matrix, the missing element, say a_{ij}, can be substituted by w_i/w_j. Fixing every problem regarding the inconsistent pair-wise comparisons by the above procedures, we can readily move on to the next step of MOSC.

C. Web-based Implementation of MOSC

This part introduces the implementation of MOSC on the Internet as a client–server architecture[8] to carry out MOP readily and effectively [33, 35]. The core of the system is divided into a few independent modules each of which

---

[8] http://www.sc.tutpse.tut.ac.jp/Research/multi.html

is realized using the appropriate implementation tools. An identifier module provides a modeling process of the value function using a neural network where a pair-wise comparison is easily performed in an interactive manner following the instructions displayed on the Web pages. An optimizer module solves a SOP under the identified value function. Moreover, a graphic module generates various graphics for illustrating outcomes.

The user interface of the system is a set of Web pages created dynamically during the solution process. The pages described in HTML (hypertext markup language) are viewed by the user's browser, which is a client of the server computer. The server computer is responsible for data management and computation, whereas the client takes care of input and output procedures. That is, users are required to request a certain service and input some parameters, and in turn, receive the result through visual and/or sensible browser operation as illustrated in Figure 3.16. In practice, the user interface is a program creating HTML pages and transferring information between the client and the server.

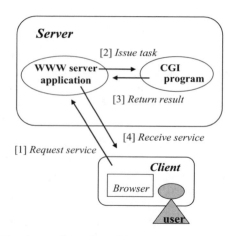

Fig. 3.16. Scheme of task flow through CGI

The HTML pages are programmed using common gateway interface (CGI) programming languages such as Perl and/or Ruby. As is the role of CGI, every job is executed on the server side and no particular tasks are assigned to the browser side. Consequently, users are not only free from the maintenance of the system but also unconstrained in their computation environment, like operating system, configuration, performance, *etc.*

Though there are several Web sites serving the (single-objective) optimization library[9], none is known except for NIMBUS [43][10] regarding MOP.

---

[9] *e.g.*, http://www-neos.mcs.anl.gov/
[10] http://nimbus.math.jyu.fi/

Since the method employed in NIMBUS is interactive, it has some stiffness as mentioned in Appendix C.

D. Integration of Multi-objective Optimization with Modeling

Usually, earlier stages of the product design task concern a model building that aims at revealing a certain relation between requirements or performances and design variables correctly. To add the value while keeping specification of the product is the designer's chance to show his/her ability. Under the diversified customer demands, the decision on what are key issues for competitive product development is strongly dependent on the designer's sense of value. Eventually, it may refer to an intent structure of the designers or a set of attributes of the performance and the preference relation among them. In other words, we need to model them as a value function at the next stage of the design task. Finally, how much we can do well depends greatly on the success in modeling of the value function.

In addition to the usual physical model-based approaches in product design/development, certain simulation-based approaches often take place by virtue of the outstanding progress of the associated technologies, *i.e.*, high performance computers, sophisticated simulation tools, novel information technologies, *etc.* These technologies are in the process of bringing about a drastic reduction in lead-time and human load in engineering tasks through rapid inspection and evaluation of products. They are trying to replace certain time-consuming and/or labor-intensive tasks like prototyping, evaluation and improvement with an integrated intelligence in terms of computer-aided simulation, analyses and syntheses.

Particularly, if designers are engaged in multi-objective decisions, they are required to repeat a process known as the P(lan)-D(o)-C(heck)-A(ct) cycle many times before attaining a final goal. As depicted in the upper part of Figure 3.17, even if they adopt the simulation-based approach, it might require a considerable load to attain the final goal especially for complicated and large-scale problems. Therefore, to cope with such a situation practically is becoming of increasing interest. For example, the response surface method [44] has been widely applied for SOP. It tries to attain a satisfactory and highly reliable goal while spending fewer effort to create a response surface in the aid of design of experiment (DOE) . The DOE is a useful technique for generating response surface models from the execution results. DOE can encourage the use of designed simulation where the input parameters are varied in a carefully structured pattern to maximize the information extracted from the resulting simulations or output results.

Though various techniques for mapping these input–output relations are known, the RBF network used for value function modeling is adequate, since we are concerned with the problem using the common technique. This kind of approach is said to be a meta-model-base since the decision will be made based on the model derived from a set of analyses given by another model, *e.g.*, finite

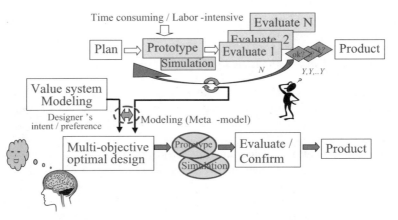

Fig. 3.17. Comparison between conventional and agile system developments

element model (FEM), regression model, *etc.* As illustrated in the lower part of Figure 3.17, decision support with this scheme can be expected to drastically reduce the lead time and effort required for product development toward agile manufacturing based on flexible integrated product design optimization.

Associated with the multi-objective design, this approach becomes much more favorable if the value system of a designer as a DM can be modeled in a cooperative manner with the meta-modeling process. In doing so, it should be noticed that the validity of the simulation is limited within a narrow space concerned for various reasons. At the early stages of the design task, however, it is quite difficult or troublesome to set up such a specified design space that is close enough, or equivalently, precise enough to describe the system around the final design which is unknown at this stage. Consequently, if the resulting design is far from what the designer prefers, further steps should be directed towards the improvement of both models, *i.e.*, the design model and the value system model. Though increasing the sampling points for the modeling is a first thought to cope with such problem, it expands the load of responses in value function modeling and consumes much computation time in the meta-modeling. On the other hand, even under the same number of sampling points, we can derive a more precise and relevant model if we narrow the modeling space. However, this may cause such a risk that the truly desired solution may be missed since it could lie outside the modeling space. In such dilemma, a promising approach is to provide a progressive procedure by interrelating the value function modeling to the meta-modeling. Beginning with building a rough model for each, the approach is intended to attain the preferentially optimal solution gradually through updating both models along with the path that will guide the tentative solution to the optimal one. Such an approach may improve the complex and complicated design process while reducing the designer's load to express his/her preference and to achieve his/her goal.

As a rough modeling technique of the value function suitable at the first stage, the following procedure is appropriate from a certain engineering sense. After setting up the utopia and nadir of each objective function, ask the DM to reply his/her upper and lower aspiration levels instead of the pair-wise comparison procedure stated in Sect. 3.3.1. Such responses seem easier for designers compared with the pair-wise comparison on the basis of objective values. This is because the designer always conceives his/her reference values when engaging in the design task. In practice, this will be done as follows. Let us define the upper aspiration level f^{UAL} as the degree to be "very" superior to the nadir or "somewhat" inferior to the utopia, and the lower aspiration level f^{UAL} to be "fairly" superior to the nadir, or "pretty" inferior to the utopia. Then, ask the DM to answer these values for every objective by setting up appropriate standards. For example, define the upper aspiration level as the point 20% inferior to the utopia or 80% superior to the nadir, and the lower aspiration level 30% superior to the nadir, or 50% inferior to the utopia. Results of the responses are transformed automatically by each element of the predetermined PWCM as shown in Table 3.2. Being free from the pair-wise comparison that may be a bit tedious for the DM, we can reduce the load of the DM in the value function modeling at the first step.

Table 3.2. Pair-wise comparison matrix (primary stage)

	f^{utop}	f^{UAL}	f^{LAL}	f^{nad}
f^{utop}	1	3	7	9
f^{UAL}	1/3	1	5	7
f^{LAL}	1/7	1/5	1	3
f^{nad}	1/9	1/7	1/3	1

Equally: 1, Moderately: 3, Strongly: 5,
Demonstrably: 7, Extremely: 9

Since the first tentative solution resulting from the thus identified value function and the rough meta-model is generally unsatisfactory for the DM, a certain iterative procedure should be taken to improve the quality of the solution. First the meta-model will be updated by adding new data near the tentative solution and deleting old data far from it. Under the expectation that the tentative solution tends gradually to the true optimum, some records of the search process in the optimization provide useful information[11] for the selection of new sampling data for the meta-modeling.

Supposing that the search process moves along the trajectory like $\{x^1, x^2, \ldots, x^k, \ldots, \hat{x}^*\}$, the direction $d_k = \hat{x}^* - x^k$ corresponds to a rough descent direction to the optimal point in the search space. Preparing two hyper spheres centered at the tentative solution and with the different diameters as

---

[11] This idea is similar to that of long-term memory in tabu search.

illustrated in Figure 3.18, it makes sense to delete the data outside the larger sphere and to add some data on the surface of the smaller sphere besides the tentative solution $\hat{x}^*$,

$$x^k_{add} = \hat{x}^* + \text{rand(sign)} r \cdot d_k / \parallel d_k \parallel \quad (k = 1, 2, \ldots),$$

where r denotes the diameter of the smaller sphere and rand(sign) randomly takes a positive or a negative sign.

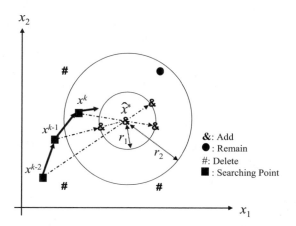

Fig. 3.18. Renewal policy using the foregoing searching process

According to the rebuilt meta-model, the foregoing value function should also be updated around the tentative solution. Additional points should be chosen due to the fact that the pair-wise comparison between too close points makes the subjective judgment difficult. After collecting the preference information from the pair-wise comparison between the remaining trials and the additional ones, a revised value function is obtained through relearning of the RBF network. By replacing the current models with the revised models, in turn, the problems will be solved repeatedly until a satisfactory solution is obtained. In the value function, f^R is initially set at the center of the search space and at the tentative solution after that.

In summary, the design optimization procedure presented here makes it possible to carry out MOP regardless of the nature of model, *i.e.*, whether it is a physical model or a meta-model. After the model selection, the next step is merged into the same flow. To restrict the search space and the modeling extent of the value function as well, the utopia and nadir solutions are to be set forth in the objective function space. Within a thus prescribed space, several trial solutions are generated properly. Then, ask the DM to perform the pair-wise comparisons, or assign the under- and lower-aspiration levels mentioned already. If the consistency of the pair-wise comparisons is satisfied (the

PWCM in Table 3.2 is consistent), they are used to train the neural network and to identify the value function. If not, fix the consistency problems based on the methods presented already. Finally, by applying an appropriate optimization method, the tentative solution is derived. If the DM accepts it, stop. Otherwise, repeat the adequate procedure depending on the circumstances until a satisfied solution is obtained.

3.4 Applications of MOSC for Manufacturing Optimization

Multi-objective optimization (MOP) has received increasing interest as a decision aid supporting agile and flexible manufacturing. To facilitate the wide application of MOP in complex and global decision environments under the manifold sense of value, applications of MOSC ranging from a strategic planning to an operational scheduling are presented below.

The location problem of a hazardous waste disposal site is an eligible interest associated with environmental and economic concerns. From such an aspect, a site location problem of hazardous waste is shown first. The second topic concerns a multi-objective scheduling optimization that has been increasingly considered an important problem-solving in manufacturing planning. Though several formulations have been proposed as mathematical programming problems, few solution methods have been found for the multi-objectives due to the special complexity of the problem class. Against this, the suitability of the MOSC approach will be shown. Thirdly, a multi-objective design optimization will be illustrated by taking a simple artificial product design first, and extending it to the integrated optimization of value function modeling and meta-modeling. Here, meta-model means the model that maps independent variables to dependent ones after these relations have been revealed by using another model.

Because of the generic property of MOP mentioned already (subjective decision problem), it is impossible to derive a preferentially optimal solution by the mathematical conditions only. To verify the effectiveness of the method throughout the following applications, therefore, we suppose the common virtual DM whose preference will be given as a utility function defined by

$$U(f(x)) = \left[\sum_{i=1}^{N} w_i \left\{ \frac{f_i^{\mathrm{nad}} - f_i(x)}{f_i^{\mathrm{nad}} - f_i^{\mathrm{utop}}} \right\}^p \right]^{1/p} \quad (p = 1, 2, \ldots), \tag{3.11}$$

where w_i denotes a weighting factor, p a parameter to specify the adopted norm[12], and f_i^{utop} and f_i^{nad} utopia and nadir values, respectively.

---

[12] (1) linear norm ($p = 1$), (2) squared norm ($p = 2$), and (3) min-max norm ($p = \infty$) are well-known.

Moreover, to simulate the virtual DM's preference, *i.e.*, subjective judgment in the pair-wise comparisons, the degree of preference already mentioned in Table 3.1 is assumed to be given by

$$
\begin{cases}
a_{ij} = 1 + \left[\frac{8(U(f^j) - U(f^i))}{U(f^{\mathrm{nad}}) - U(f^{\mathrm{utop}})} + 0.5\right], & \text{if } U(f^i) \geq U(f^j) \\
a_{ij} = 1/a_{ji}, & \text{otherwise}
\end{cases}
\tag{3.12}
$$

where $[\cdot]$ denotes the Gauss symbol. This equation gives $a_{\mathrm{utop,nad}} = 9$ for such a statement that the utopia is extremely favorable to the nadir. Also when $i = j$, it returns, $a_{ii} = 1$. By comparing the result obtained from the MOSC to the reference solution that will be derived from the direct optimization under Equation 3.11, *i.e.*, Problem 3.13, it is possible to verify the effectiveness of the approach,

$$
[Problem] \quad \max \ U(f(x)) \text{ subject to } x \in X.
\tag{3.13}
$$

3.4.1 Multi-objective Site Location of Waste Disposal Facilities

Developing a practical method of location problem for hazardous waste disposal [31] is meaningful as a key issue in considering a sustainable technology under environmental and economic concerns.

A basic but general formulation of the location problem of the disposal site shown in Figure 3.19 is described such that: for rational disposal of the hazardous waste generated at L sources, choose the suitable sites up to K among the M candidates. The objective functions are composed of cost and risk, and decision variables involve real variables giving the amount of waste shipped from source to site, and 0-1 variables each of which takes 1 if the site is open and 0 otherwise.

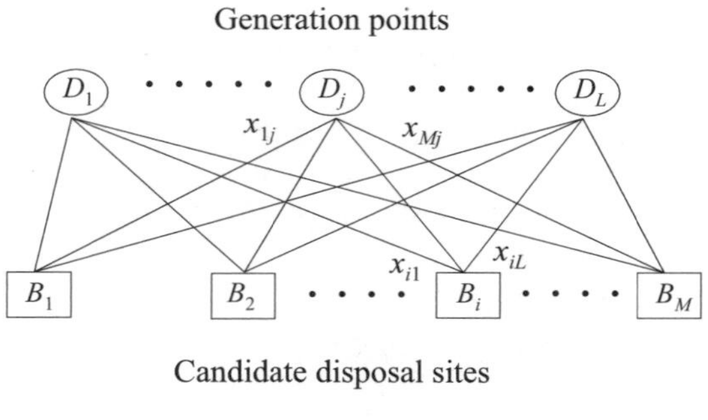

Fig. 3.19. A typical site location problem

Since the conflict between economy and risk is common to this kind of NIMBY (not in my back yard) problem, this problem can be described adequately as a bi-objective mixed-integer linear program (MILP) ,

$$[Problem] \min_{x,z} \ \{f_1 = \sum_{i=1}^{M}\sum_{j=1}^{L} C_{ij}x_{ij} + \sum_{i=1}^{M} F_i z_i,$$

$$f_2 = \sum_{i=1}^{M}\sum_{j=1}^{L} R_{ij}x_{ij} + \sum_{i=1}^{M} Q_i B_i z_i\},$$

$$\text{subject to} \ \begin{cases} \sum_{i=1}^{M} x_{ij} \geq D_j \ (j=1,\ldots,L) \\ \sum_{j=1}^{L} x_{ij} \leq B_i z_i \ (i=1,\ldots,M) \\ \sum_{i=1}^{M} z_i \leq K. \end{cases} \quad (3.14)$$

In the above, f_1 and f_2 denote the objective functions evaluating cost and risk, respectively. They are functions of the amount of waste shipped from source j to site i, $x_{ij}(\geq 0)$, and 0-1 variable $z_i(\in \{0,1\})$, which takes 1 if the i-th site is chosen and 0 otherwise. Moreover, D_j denotes demand at the j-th source and B_i capacity at the i-th site. Then, the first condition of Equation 3.14 describes that the waste is shippable at each source, and the second one is disposable at each site. Moreover, K is an upper bound of the allowable construction of the site.

On the other hand, C_{ij} denotes the shipping cost from j to i per unit amount of waste, and F_i the fixed-charge cost of site i. R_{ij} denotes the risk constant accompanying transportation per unit amount from j to i. Generally, it may be a function of distance, population density along the traveling route, and other specific factors. Likewise, Q_i represents the fixed-portion of risk at the i-th site per unit capacity; it is considered to be a function of population density around the site, and some other specific factors.

The above problem is possible to solve by the MOHybGA mentioned in Sect. 3.3.2 after reformulation in a hierarchical manner,

$$[Problem] \quad \max_{z,\epsilon_2} V_{NN}(f_1(x,z),\epsilon_2; f^R) - P \cdot \max[0, \sum_{i=1}^{M} z_i - K],$$

$$\text{subject to} \ \min_{x} f_1(x,z)$$

$$\text{subject to} \ \begin{cases} f_2(x,z) \leq f_2^* + \epsilon_2 \\ \sum_{i=1}^{M} x_{ij} \geq D_j \ (j=1,\ldots,L) \\ \sum_{j=1}^{L} x_{ij} \leq B_i z_i \ (i=1,\ldots,M) \end{cases}.$$

The pure constraint on integer variables is handled by a penalty term in the objective function at the master problem where P denotes a penalty coefficient, and $\max[\cdot]$ is the operator returning the greatest among the arguments.

Since the system equations and two objective functions are all linear functions of the decision variables, it is easy to solve the slave problem using linear programming even if the problem size may become very large.

Numerical experiments take place for the problem where $M = 8$, $L = 6$ and $K = 3$. Parameters of GA are set as crossover rate $= 0.1$, mutation rate $= 0.01$, and population size $= 50$ for the chromosome 11 bits long.

A virtual DM featured in Equations 3.11 and 3.12 evaluates the preference among five trial solutions (B, C, D, E, F), shown in Figure 3.21. Using the value function modeled by the PWCM in Figure 3.20, a best compromise solution is obtained after 14 generations. In Figure 3.21, the POS set is imposed on a set of contours of value function. The best compromise solution is obtained at point A, which locates on the POS set and has the highest value of the value function at the same time.

	f^1	f^2	f^3	utopia	nadir
f^1	1	3	1/3	1/5	5
f^2		1	1/5	1/6	3
f^3			1	1/3	5
utopia	$a_{ji} = 1/a_{ij}$			1	9
nadir					1

Fig. 3.20. Pay-off matrix for the site location problem

3.4.2 Multi-objective Scheduling of Flow Shop

The multi-objective scheduling has received increasing attention as an important problem-solving method in manufacturing. However, optimization of production scheduling refers to integer and/or mixed-integer programming problems whose combinatorial nature makes the solution process very complicated and time-consuming (*NP*-hard). Since its multi-objective optimization will amplify the difficulty, it has scarcely been studied previously [45].

Among others, Murata, Ishibuchi and Tanaka [46] recently studied a flow shop problem under two objectives such as makespan and total tardiness using a multi-objective genetic algorithm (MOGA). In Bogchi's book [47], flow shop, open shop and job shop problems were discussed under the two objectives, makespan and average holding time. Bogchi applied NSGA and a elitist non-dominated sorting genetic algorithm (ENGA). Moreover, Saym and Karabau [48] used a branch and bound method for a similar kind of problem. A parallel machine problem was solved by Tamaki, Nishino and Abe [49] under consideration of total holding time and discrepancy from due

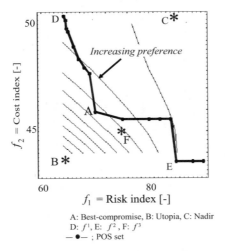

A: Best-compromise, B: Utopia, C: Nadir
D: f^1, E: f^2, F: f^3
— •— ; POS set

Fig. 3.21. Best compromise solution for the site location problem

date using parallel selection Pareto reserve GA (PPGA) and also by Mohri, Masuda and Ishii [50] so as to minimize the maximum completion time and maximum lateness. On the other hands, Sakawa and Kubota [51] studied the job shop problem under three fuzzy objectives by multiple-deme GA and a multi-objective tabu search was applied for a single-machine problem with sequence-dependent setup times [52]. However, these studies only derived the POS set that presents a bulk of candidates of the final solution.

In what follows, MOON^{2R} is applied to derive the preferentially optimal solution for a two-objective flow shop scheduling problem. Under the mild assumptions that no jobs are dividable, simultaneous operations are inhibited on machines and every processing time and due date are given, the problem is formulated. The goal of this problem is to minimize the sum delay of due time f_1 and the total changeover cost f_2. The scheduling data is generated randomly within certain extents, *i.e.*, between 1 and 10 for due time and every four intervals between 4 and 40 for changeover cost, respectively.

Among the trial solutions generated as shown in Figure 3.22, PWCM of the virtual DM is given as Table 3.3 based on Equations 3.11 and 3.12 ($p = 1$, $w_1 = 0.3$, $w_2 = 0.7$). The total number of responses becomes 35 in this case ($a_{\text{utop,nad}} = 9$ is implied).

In Figure 3.23, the contours of preference (indifference curves) are compared with those of the presumed ones and V_{RBF} (f, f^R)[13] when $p = 1$. Except for the marginal regions, the identified (solid curve) and the original (dotted line) almost coincide with each other.

---

[13] Presently, f^R is set at $(0,0)$.

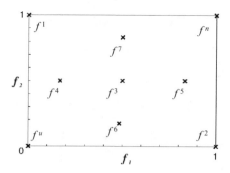

Fig. 3.22. Location of trail solutions

Table 3.3. Pair-wise comparison matrix ($p = 1$)

	f^u	f^n	f^1	f^2	f^3	f^4	f^5	f^6	f^7
f^u	1	9	3	7	5	3	7	4	6
f^n	1/9	1	1/7	1/3	1/5	1/7	1/3	1/6	1/4
f^1	1/3	7	1	4	3	1	4	2	3
f^2	1/7	3	1/4	1	1/3	1/4	1	1/3	1
f^3	1/5	5	1/3	3	1	1/3	3	1/2	2
f^4	1/3	7	1	4	3	1	5	2	4
f^5	1/7	3	1/4	1	1/3	1/5	1	1/4	1/2
f^6	1/4	6	1/2	3	2	1/2	4	1	3
f^7	1/6	4	1/3	2	1/2	1/4	2	1/3	1

With the thus identified value function, the following three flow shop scheduling problems are solved by MOON^{2R} with SA as the optimization technique:

1. One process, one machine and seven jobs
2. Two processes, one machine and ten jobs
3. Two processes, two machines and ten jobs.

The SA employed the conditions that the insertion neighborhood[14] is adopted, reduction rate of the temperature $= 0.95$, and number of iterations $= 400$.

Table 3.4 summarizes numerical results in comparison with the reference solutions. It is known that MOON^{2R} can derive the same results in every case ($p = 1$). Figure 3.24 is a Gantt chart showing a visual examination of the feasibility of the result.

As the number of trial solutions is decreased gradually from the foregoing 9 to 7 and 5, the number of required responses of the DM will decrease until 20

---

[14] A randomly selected symbol is inserted into a randomly selected position, *e.g.*, $A - (B) - C - D - (\cdot) - E - F$ is changed into $A - C - D - B - E - F$ if the parentheses denote the random selections.

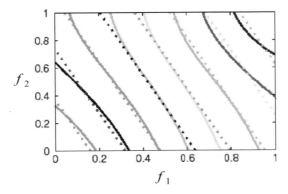

Fig. 3.23. Comparison of contours of value functions ($p = 1$).

Table 3.4. Comparison of numerical results($p = 1,\ 2$)

Type of problem	Type of value function			
	$p=1$		$p=2$	
	Reference	V_{RBF}	Reference	V_{RBF}
(1,1, 7) *	3.47	3.47	1.40	1.40
(2,1,10)	7.95	7.95	2.92	2.92
(2,2,10)	3.40	3.40	1.60	1.60

* Number of process, machine, and job.

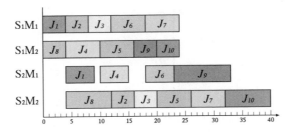

Fig. 3.24. Gantt chart of the (2,2,10) problem ($p = 1$).

and 10, respectively. The last number is small enough for the DM to respond acceptably. Every case derived the same result as shown in Table 3.4. This means the linear value function can be identified correctly with a small load of interaction.

In the same way, the case of the quadratic form of the value function is solved successfully as shown both in Figure 3.25 and Table 3.4 ($p = 2$). Due to the good approximation of the value function (the identified: solid curve, the original: broken line), MOSC can also derive the same results as the reference.

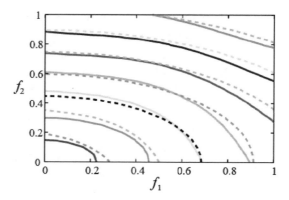

Fig. 3.25. Comparison of contours of value functions ($p = 2$)

3.4.3 Artificial Product Design

A. Design of a Beam Structure

Here, we show the results of applying MOON^2 to the beam structure design problem as formulated below,

$$[\textit{Problem}] \quad \min \ f(x) = \{f_1(x), f_2(x)\}$$

$$\text{subject to} \begin{cases} g_1(x) = 180 - \frac{9.78 \times 10^6 x_1}{4.096 \times 10^7 - x_2{}^4} \geq 0 \\ g_2(x) = 75.2 - x_2 \geq 0 \\ g_3(x) = x_2 - 40 \geq 0 \\ g_4(x) = x_1 \geq 0 \\ h_1(x) = x_1 - 5x_2 = 0 \end{cases} \quad , \quad (3.15)$$

where x_1 and x_2 denote the tip length of the beam and the interior diameter, respectively, as shown in Figure 3.26. Inequality and equality equations represent the design conditions. Moreover, objective functions f_1 and f_2 represent the volume (equivalently, weight) of the beam [mm^3] and static compliance of the beam [mm/N], respectively. These are described as follows:

$$f_1(x) = \frac{\pi}{4} \left[x_1 \left(D_2{}^2 - x_2{}^2 \right) + (l - x_1) \left(D_1{}^2 - x_2{}^2 \right) \right],$$

$$f_2(x) = \frac{64}{3\pi E} \times \left[\left(\frac{1}{D_2{}^4 - x_2{}^4} - \frac{1}{D_1{}^4 - x_1{}^4} \right) x_1{}^3 + \frac{l^3}{D_1{}^4 - x_1{}^4} \right],$$

where E denotes Young's modulus. There is a tradeoff such that the smaller static compliance needs the tougher structure (larger volume), and *vice versa*.

Figure 3.27 shows the locations of the trial solutions generated based on Equation 3.4. The pair-wise comparison matrix of the virtual DMDM!virtual

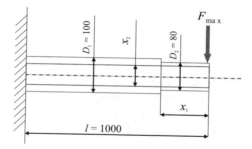

Fig. 3.26. Beam structure design problem

is given in Table 3.5 for $p = 1$. Omitting some data left for the cross validation (shown in italics in the table), these are used to model the value function by a BP neural network with ten hidden nodes. Both inputs and an output of the neural network are normalized between 0 and 1. Then, the original problem is rewritten as follows:

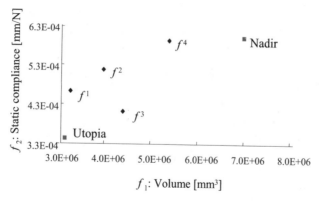

Fig. 3.27. Generated trial solutions (linear)

[*Problem*]   max $V_{NN}(f_1(x), f_2(x), f^R)$ subject to Equation 3.15.

To solve the above problem, the sequential quadratic programming (SQP) is applied with the numerical differentiation described as Equation 3.9.

Numerical results for three cases, *i.e.*, $p = 1, 2$, and ∞ are shown in Figure 3.28 and Table 3.6. From the figures, where contours of value function are superimposed, it is known in every case that:

1. The shape of the value function is described properly.
2. The solution ("By MOON²") locates on the Pareto optimal set and it is almost identical to the reference solution ("By utility function").

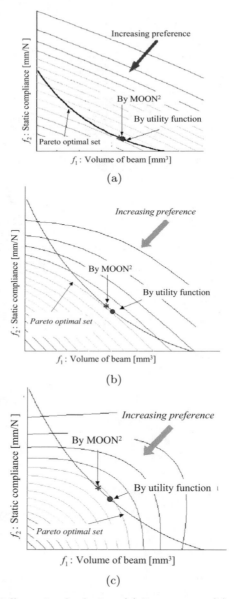

Fig. 3.28. Preferentially optimal solution: (a) linear norm, (b) quadratic norm, (c) min-max norm

The results summarized in the table include the weighting factor w, inconsistency index CI, and root mean squared errors e at the training and

Table 3.5. Pair-wise comparison matrix $(p = 1)$

	f^{utop}	f^{nad}	f^1	f^2	f^3	f^4
f^{utop}	1.0	9.0	3.67	5.32	3.28	7.82
f^{nad}	0.11	1.0	0.16	0.21	0.15	*0.46*
f^1	0.27	6.33	1.0	*2.65*	0.72	5.14
f^2	0.19	4.68	0.38	1.0	0.33	3.49
f^3	0.3	6.72	1.39	3.04	1.0	5.53
f^4	0.13	*2.18*	0.19	0.29	0.18	1.0

Table 3.6. Summary of results

Type		Reference	MOON^2
Linear $(p = 1)$	f_1	5.15E+6	5.11E+6
$w=(0.3,0.7)$	f_2	3.62E-4	3.63E-4
$CI = 0.051$	x_1	251.4	253.6
$e =(1.8E-3,3.5E-2)$	x_2	50.3	50.7
Quadratic $(p = 2)$	f_1	4.68E+6	4.56E+6
$w=(0.3,0.7)$	f_2	3.77E-4	3.82E-4
$CI = 0.048$	x_1	275.7	281.5
$e =(1.3E-2,1.4E-1)$	x_2	55.1	56.3
Min-Max $(p = \infty)$	f_1	4.5E+6	4.3E+6
$w=(0.4,0.6)$	f_2	3.8E-4	3.9E-4
$CI = 0.019$	x_1	283.4	292.9
$e =(6.7E-3,5.3E-2)$	x_2	56.7	58.6

validation stages of the neural network. It is known that satisfactory results are obtained for every case.

Except for in the linear case, however, there is a little room left for improvement. Since the utility functions become more complex in the order of linear, quadratic, and min-max form, modeling of the value functions becomes more difficult in the same order. This causes distortion of the value function everywhere where the evaluation is far from the fixed input f^R. Generally, it is hard to attain the rigid solution only by the search performed within the global space. To obtain the more correct solution, it is necessary to limit the search space around the earlier solution and repeat the same procedure as described in the flow chart in Figure 3.14.

B. Design of a Flat Spring Using a Meta-model

Let us consider the design problem of a flat spring as shown in Figure 3.29. The aim of this problem is to decide the shape of spring (x_1, x_2, x_3) so as to increase the rigidity f_2 while reducing the stress f_1. Because it is impossible to achieve these objectives at the same time, it is amenable to formulating the problem as MOP.

Generally, as the shape of a product becomes complicated, it becomes accordingly hard to model mathematically the design objectives with respect

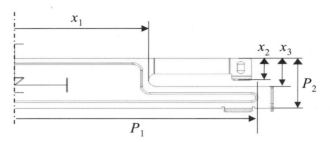

Fig. 3.29. Flat spring design

Table 3.7. Design variables and design objectives

x_1	x_2	x_3	f_1(stress)	f_2(rigidity)
0.0	0.0	0.0	0.0529	0.0000
0.0	0.5	0.6	0.0169	0.0207
0.0	1.0	1.0	0.0000	0.0322
0.5	0.0	0.6	0.1199	0.1452
0.5	0.5	1.0	0.0763	0.1462
0.5	1.0	0.0	0.9234	0.3927
1.0	0.0	1.0	0.2720	0.4224
1.0	0.5	0.0	1.0000	1.0000
1.0	1.0	0.6	0.5813	0.8401

(Normalized between 0 and 1)

to design variables. Under such circumstances, computer simulation methods such as the finite element method (FEM) has been widely used to reveal the relation between them. In such a simulation-based approach, DOE also plays an important role. Presently, three levels of the designed experiments are set for each design variable. Then two design objectives are evaluated by a set of design variable values derived from the orthogonal design of DOE. Results of the simulation from the FEM model are then used to derive a model that can explain the relation or to construct the response surface. For this purpose, meta-models of f_1 and f_2 with respect to (x_1, x_2, x_3) are derived by using an RBF network that uses the FEM results shown in Table 3.7. For the sake of convenience, the results of such modeling will be represented as Meta-$f_1(x_1, x_2, x_3)$ and Meta-$f_2(x_1, x_2, x_3)$.

On the other hand, to reveal the value function of the DM, the utopia and nadir are set, respectively, at $(0, 0)$ and $(1, 1)$ after normalizing the objective value on a basis of unreachability[15]. Within the space surrounded by these two points in the objective space, seven trial solutions are generated randomly, and the imaginary pair-wise comparison is performed as before under the condition that $p = 1, w_1 = 0.4, w_2 = 0.6$. Then, the RBF network is trained to derive the value function as V_{RBF} (Meta-f_1, Meta-f_2), by which we can evaluate the

---

[15] Hence, 0 corresponds to utopia, and 1 to nadir.

Table 3.8. Comparison between two methods

		Reference	MOON^{2R}	
			1st	2nd
Design	x_1 [mm]	43.96	43.96	43.96
variable	x_2 [mm]	5.65	5.92	5.95
	x_3 [mm]	9.11	9.65	9.10
Objective	f_1 [MPa]	1042.70	867.89	964.795
function	f_2 [N/mm]	9.05	8.23	8.57

objective functions for the arbitrary decision variables. Now the problem can be described as follows:

$$[Problem] \quad \max \ V_{\mathrm{RBF}}(\text{Meta-}f_1(x), \text{Meta-}f_2(x))$$
$$\text{subject to} \begin{cases} 0 \le x_1 \le P_1 \\ 0 \le x_2 \le x_3 \ , \\ 0 \le x_3 \le P_2 \end{cases} \quad (3.16)$$

where Meta-$f_1(x)$ and Meta-$f_2(x)$ denote the meta-model of the stress and the rigidity, respectively, and P_1 and P_2 denote the parameters specific to the shape of the spring. The resulting optimization problem is solved using the revised simplex method (refer to Appendix B) that can handle the upper and lower bound constraints.

By comparing the results between the columns named "Reference" and "MOON^{2R} (1st)" in Table 3.8, "MOON^{2R}" solution is shown to be very close to the reference solution in the decision variable space. In contrast, there exists some discrepancy between the results in the objective function space (especially regarding f_1). This is because f_1 is very sensitive with respect to the design variables around the (tentative) optimal point. By supposing that the DM would not be satisfied with the result, let us move on and revise the tentative solution in the next step. After shrinking the search space around the tentative solution, the new utopia and nadir are set at (0.03, 0.60) and (0.25, 0.818), respectively. Then the same procedures are repeated, *i.e.*, generate five trial solutions, perform the pair-wise comparison, and so on. Thereafter, a new result is obtained as shown in "MOON^{2R} (2nd)" in Table 3.8. The foregoing results are known to be updated quite well. If the DM feels that there still remains some room for improvement in the result, further steps are necessary. Such action gives rise to additional procedures to correct the meta-model around the tentative solution. Presently since both meta-model and value function are given by the RBF network, we can use the increment operations of RBF network to save the computation loads for these revisions.

C. Design of a Beam through Integration with Meta-modeling

The integrated approach through inter-related modeling of the value system and the meta-model will be illustrated by reconsidering the beam design prob-

Table 3.9. Results of FEM analyses

	No.	x_1 [mm]	x_2 [mm]	C_{st} [mm/N]
	1*	10	40	0.000341
	2*	10	57.5	0.000375
	3*	10	75	0.000490
	4*	255	40	0.000351
Primal	5	255	57.5	0.000388
	6	255	75	0.000550
	7*	500	40	0.000408
	8	500	57.5	0.000469
	9	500	75	0.000891
	10	320.57	64.11	0.000438
Addi-	11	337.05	67.41	0.000472
tional	12	274.43	65.29	0.000433
	13	366.72	62.94	0.000445

lem [53]. However, this time it is assumed that the static compliance of the beam is available only as a meta-model. By denoting such a meta-model of the compliance as Meta-$f_2(x)$, the design problem is described as follows:

$$[Problem] \quad \min \quad \{f_1(x) = \tfrac{\pi}{4}(x_1(D_2^2 - x_2^2) + (l - x_1)(D_1^2 - x_2^2)), \text{Meta-}f_2(x)\}$$

$$\text{subject to} \quad \begin{cases} g_1 = 180 - \max(\frac{3.84\times10^{10}}{\pi(100^4 - x_2^4)}, \frac{3.072\times10^7 x_1}{\pi(80^4 - x_2^4)}) \geq 0 \\ g_2 = 75.2 - x_2 \geq 0 \\ g_3 = x_2 - 40 \geq 0 \\ g_4 = x_1 \geq 0 \\ h_1 = x_1 - 5x_2 = 0 \end{cases}$$

Table 3.10. Primal references and additional trial

		f_1 [mm^3]	f_2 [mm/N]
	f^{uto}	2.02 ×10^6	3.38 ×10^{-4}
Primal	f^{UAL}	3.16 ×10^6	4.70 ×10^{-4}
	f^{LAL}	5.43 ×10^6	7.33 ×10^{-4}
	f^{nad}	6.57 ×10^6	8.64 ×10^{-4}
Additional	f^1	3.71 ×10^6	4.45 ×10^{-4}

To have the meta-model, the FEM analysis is carried out for every pair of three levels of x_1 and x_2. Then using the results x_1, x_2, and C_{st} listed in Table 3.9 (primal), the primal meta-model is derived as a RBF network model, *i.e.*, $(x_1, x_2) \to C_{st}$. On the other hand, the primal value function is obtained from the preference information that will not depend on the pair-wise comparison and use the references. That is, data shown in Table 3.10 (primal) is adopted as the reference values.

Table 3.11. Comparison of the results after compromising

		x_1 [mm]	x_2 [mm]	f_1 [mm^3]	f_2 [mm/N]
Reference		343.58	68.72	3.17 $\times 10^6$	4.79$\times 10^{-4}$
MOON^{2R}	1st	3.21 $\times 10^6$	64.11	3.72 $\times 10^6$	4.66 $\times 10^{-4}$
	2nd	3.44$\times 10^6$	68.72	3.17 $\times 10^{-4}$	4.89$\times 10^{-4}$ (Meta-f_2)

Following the procedures mentioned already, the primal solution is obtained as shown by "1st" in Table 3.11 by applying SQP as the optimization method.

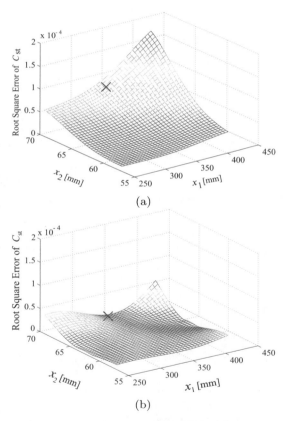

Fig. 3.30. Error of response surface near the target: (a) 1st stage, (b) 2nd stage

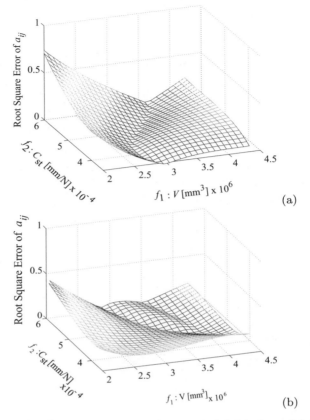

(a)

(b)

Fig. 3.31. Value function error near the target: (a) 1st stage, (b) 2nd stage

Such a primal solution would not normally satisfy the DM (actually there are discrepancies between by "1st" and "reference" solutions.). The next step to update the solution is taken by rebuilding both primal models. For meta-model rebuilding, the data marked by asterisks (1, 2, 3, 4, 7) are deleted and four data are augmented as shown in Table 3.9. Meanwhile, the value function is modified by adding the data shown in Table 3.10 and this requires the DM to make the pair-wise comparisons between the added and the existing data. (The value function of the virtual DM is prescribed as before with parameters like $p = 1$, $w_1 = 0.4$, $w_2 = 0.6$). Errors in the course of model building are compared in Figure 3.30 for the meta-model, and Figure 3.31 for the value function, respectively. From these results, the simultaneous improvement is achieved by the integrated approach.

3.5 Chapter Summary

To deal with diversified customer demands and global competition, requirements on agile and flexible manufacturing are being continuously increased. Multi-objective optimization (MOP) has accordingly been taken as a suitable decision aid supporting such circumstances. This chapter focused on recent topics associated with multi-objective problems.

First, the extended applications of evolutionary algorithm (EA) were presented as a powerful method associated with the multi-objective analysis. Since every EA considers the multiple possible solutions simultaneously in the search, it can favorably generate the POS set in a single run of the algorithm. In addition, since it is insensitive to the concave shape or continuity of the Pareto front, it can reveal the tradeoff relation for real world problems advantageously.

As one of the most promising methods for MOP, a few methods in terms of soft computing were explained from various viewpoints. Common to those methods, a value function modeling method using neural networks was introduced. The training data of such neural network was gathered through another pair-wise comparison that is easier for the DM than AHP.

By virtue of the identified value function, an extension of the hybrid GA was shown to solve effectively MIP under multi-objectives. Moreover, using the shared fitness of GA, this approach is amenable for solving MOP including the qualitative objectives. As the major interest in the rest of this chapter, the soft computing method termed $MOON^2$ and $MOON^{2R}$ were presented. The difference of these methods is the type of neural network employed for the value function modeling. These methods can solve MOP under a variety of preferences effectively as well as practically even for an ill-posed decision environment. Moreover, to carry out MOP readily, implementation on the Internet was shown as a client–server architecture.

At the early stages of product design, designers need to engage in model building as a step of problem definition. Modeling the value functions is also an important design task at the next stage. As a key issue for competitive product development, an approach for the integration of multi-objective optimization with the modeling both of the system and the value function was presented.

To facilitate wide application in manufacturing, a few applications ranging from strategic planning to operational scheduling were demonstrated.

First, under the two objectives the location problem of a hazardous waste disposal site was solved by the hybrid GA. The second topic concerned multi-objective scheduling optimization, which is increasingly being considered as an important problem-solving task in manufacturing. Due to the special difficulty, however, no effective solutions methods are known under multi-objectives. For such a problem, MOSC was applied successfully. Third, we illustrated multi-objective design optimization taking a simple artificial product design, and its extension for the integration of modeling and design optimization in terms of meta-modeling. Here meta-model means a model that can relate independent variables to dependent ones after these relations have been revealed by another model.

References

1. Deb K (2001) Multi-objective optimization using evolutionary algorithms. Wiley, New York
2. Fonseca CM, Fleming PJ, Zitzler E, Deb K, Thiele L (eds.) (2003) Evolutionary multi-criterion optimization. Springer, Berlin
3. Coello CAC, Aguirre, Zitzler E (eds.) (2005) Evolutionary multi-criterion optimization. Springer, Berlin
4. Obayashi S, Deb K, Poloni C, Hiroyasu T, Murata T (eds.) (2007) Evolutionary multi-criterion optimization. Springer, Berlin
5. Coello CAC (2001) A short tutorial on evolutionary multiobjective optimization. In: Zitzler E, Deb K, Thiele L, Carlos A, Coello C, Corne D (eds.) Proc. First International Conference on Evolutionary Multi-Criterion Optimization (Lecture Notes in Computer Science), pp. 21–40, Springer, Berlin
6. Fourman MP (1985) Compaction of symbolic layout using genetic algorithms. Proc. 1st International Conference on Genetic Algorithms and Their Applications, pp. 141–153. Lawrence Erlbaum Associates Inc., Hillsdale
7. Allenson R (1992) Genetic algorithms with gender for multi-function optimisation. EPCC-SS92-01. University of Edinburgh, Edinburgh
8. Ishibuchi H (1996) Multi-objective genetic local search algorithm. In: Fukuda T, Furuhashi T (eds.) Proc. 1996 IEEE International Conference on Evolutionary Computation, Nagoya, pp. 119-124
9. Hajela P, Lin CY (1992) Genetic search strategies in multicriterion optimal design. Struct Optim, 4:99–107
10. Valenzuela-Rendon M, Uresti-Charre E (1997) A non-generational genetic algorithm for multiobjective optimization. In: Back T (ed.) Proc. Seventh International Conference on Genetic Algorithms, pp. 658–665. Morgan Kaufmann Publishers Inc., San Francisco
11. Schaffer JD (1985) Multiple objective optimization with vector evaluated genetic algorithms. Proc. 1st International Conference on Genetic Algorithms and Their Applications, pp. 93-100. Lawrence Erlbaum Associates Inc., Hillsdale
12. Goldberg DE, Richardson J (1987) Genetic algorithm with sharing for multimodal function optimization. In: Grefenstette JJ (ed.) Proc. 2nd International Conference on Genetic Algorithms and Their Applications, pp. 41–49. Lawrence Erlbaum Associates Inc., Hillsdale
13. Goldberg DE (1989) Genetic algorithms in search, optimization and machine learning. Kluwer, Boston
14. Fonseca CM, Fleming PJ (1993) Genetic algorithm for multi-objective optimization: formulation, discussion and generalization. Proc. 5th International Conference on Genetic Algorithms and Their Applications, Chicago, pp. 416–423
15. Srinivas N, Deb K (1994) Multiobjective optimization using nondominated sorting in genetic algorithms. Evolutionary Computation, 2:221–248
16. Horn J, Nafpliotis N (1993) Multiobjective optimization using the niched Pareto genetic algorithm. IlliGAl Rep. 93005. University of Illinois at Urbana-Champaign

17. Deb K, Agrawal S, Pratap A, Meyarivan T (2000) A fast elitist non-dominated sorting genetic algorithm for multi-objective optimization: NSGA-II. Proc. Parallel Problem Solving from Nature VI (PPSN-VI), pp. 849–858

18. Kursawe F (1990) A variant of evolution strategies for vector optimization. In Proc. Parallel Problem Solving from Nature I (PPSN-I), pp. 193–197

19. Laumanns M, Rudolph G, Schwefel HP (1998) A spatial predator–prey approach to multi-objective optimization: a preliminary study. Proc. Parallel Problem Solving from Nature V (PPSN-V), pp. 241–249

20. Kundu S, Osyczka A (1996) The effect of genetic algorithm selection mechanisms on multicriteria optimization using the distance method. Proc. Fifth International Conference on Intelligent Systems (Reno, NV). ISCA, pp. 164–168

21. Zitzler E, Thiele L (1999) Multiobjective evolutionary algorithms: a comparative case study and the strength Pareto approach. IEEE Transactions on Evolutionary Computation, 3:257–271

22. Deb K, Goldberg DE (1991) MGA in C: a messy genetic algorithm in C. Technical Report 91008, Illinios Genetic Algorithms Laboratory (IIliGAL)

23. Knowles J, Corne D (2000) M-PAES: a memetic algorithm for multiobjective optimization. Proc. 2000 Congress on Evolutionary Computation, Piscataway, vol. 1, pp. 325–332

24. Knarr MR, Goltz MN, Lamont GB, Huang J (2003) In situ bioremediation of perchlorate-contaminated groundwater using a multi-objective parallel evolutionary algorithm. Proc. Congress on Evolutionary Computation (GEC, 2003), Piscataway, vol. 1, pp. 1604–1611

25. Zitzler E, Deb K, Thiele L (2000) Comparison of multiobjective evolutionary algorithms: empirical results. Evolutionary Computation, 8:173–195

26. Czyzak P, Jaszkiewicz AJ (1998) Pareto simulated annealing–a meta-heuristic technique for multiple-objective combinatorial optimization. Journal of Multi-criteria Decision Analysis, 7:34–47

27. Jaeggi D, Parks G, Kipouros T, Clarkson J (2005) A multi-objective tabu search algorithm for constrained optimization problems. EMO 2005, LNCS 3410, pp. 490–504

28. Rakesh A, Babu BV (2005) Non-dominated sorting differential evolution (NSDE): an extension of differential evolution for multi-objective optimization. Proc. 2nd Indian International Conference on Artificial Intelligence, pp. 1428–1443

29. Robic T, Filipic B (2005) DEMO: differential evolution for multi-objective optimization. Evolutionary Computation. In: Coello CCA et al (eds.) Proc. EMO 2005, pp. 520–533, Springer, Berlin

30. Rahimi-Vahed AR, Rabbani M, Tavakkoli-Moghaddam RT, Torabi SA, Jolai F (2007) A multi-objective scatter search for a mixed-model assembly line sequencing problem. Advanced Engineering Informatics, 21:85–99

31. Shimizu Y (1999) Multi objective optimization for site location problems through hybrid genetic algorithm with neural network. Journal of Chemical Engineering of Japan, 32:51–58

32. Shimizu Y, Kawada A (2002) Multi-objective optimization in terms of soft computing. Transactions of SICE, 38:974–980

33. Shimizu Y, Tanaka Y, Kawada A (2004) Multi-objective optimization system. MOON^2 on the Internet. Computers & Chemical Engineering, 28:821–828

34. Shimizu Y, Tanaka Y (2003) A practical method for multi-objective scheduling through soft computing approach. International Journal of JSME, Series C, 46:54–59
35. Shimizu Y, Yoo J-K, Tanaka Y (2004) Web-based application for multi-objective optimization in process systems. In: Chen B, Westerberg AW (eds.) Proc. 8th International Symposium on Computer-Aided Process Systems Engineering, Kunming, pp. 328–333, Elsevier, Amsterdam
36. Orr MJL (1996) Introduction to radial basis function networks, http://www.cns.uk/people/mark. html
37. Saaty TL (1980) The analytic hierarchy process. McGraw-Hill, New York
38. Shimizu Y (1999) Multi-objective optimization of mixed-integer programming problems through a hybrid genetic algorithm with repair operation. Transactions of ISCIE, 12:395–404 (in Japanese)
39. Shimizu Y (1999) Multi-objective optimization for mixed-integer programming problems through extending hybrid genetic algorithm with niche method. Transactions of SICE, 35:951–956 (in Japanese)
40. Shimizu Y, Yoo J-K, Tanaka Y (2006) A design support through multi-objective optimization aware of subjectivity of value system. Transactions of JSME, 72:1613–1620 (in Japanese)
41. Warfield JN (1976) Societal systems. Wiley, New York
42. Harker PT (1987) Incomplete pairwise comparisons in the analytic hierarchy process. Mathemstical Modelling, 9:837–848
43. Miettinen K, Makela MM (2000) Interactive multiobjective optimization system www-nimbus on the Internet. Computers & Operations Research, 27:709–723
44. Myers RH, Montgomery DC (2002) Response surface methodology: process and product optimization using designed experiments (2nd ed.). Wiley, New York
45. T'kindt V, Billaut JC (2002) Multicriteria scheduling: theory, models and algorithms. Springer, New York
46. Murata T, Ishibuchi H, Tanaka H (1996) Multi-objective genetic algorithm and its applications to flowshop scheduling. Computers & Industrial Engineering., 30:957–968
47. Bagchi TP (1999) Multi-objective scheduling by genetic algorithms. Kluwer, Boston
48. Saym S, Karabau S (2000) Bicriteria approach to the two-machine flow shop scheduling problem. European Journal of Operational Research, 113:393–407
49. Tamaki H, Nishino E, Abe S (1999) Modeling and genetic solution for scheduling problems with regular and non-regular objective functions. Transactions of SICE, 35:662–667 (in Japanese)
50. Mohri S, Masuda R, Ishii H (1999) Bi-criteria scheduling problem on three identical parallel machines. International Journal of Production Economics, 60:529–536
51. Sakawa M, Kubota R (2000) Fuzzy programming for multi-objective job shop scheduling with fuzzy processing time and fuzzy due date through genetic algorithms. European Journal of Operational Research, 120:393–407
52. Choobineh FF, Mohebbi E, Khoo H (2006) A multi-objective tabu search for a single-machine scheduling problem with sequence-dependent setup times. European Journal of Operational Research, 175:318–337
53. Shimizu Y, Miura K, Yoo J-K, Tanaka Y (2005) A progressive approach for multi-objective design through inter-related modeling of value system and meta-model. Transactions of JSME, 71:296–303 (in Japanese)

4

Cellular Neural Networks in Intelligent Sensing and Diagnosis

4.1 The Cellular Neural Network as an Associative Memory

Computers invented in 21st century are now essential not only for industrial technology but for our daily lives. The neumann type of computers used widely at present are able to process mass information rapidly. These computers read instructions from a memory store and execute them in a central processing unit at high speed. This is why computers are superior to human beings in fields like numerical computations.

However, even if a computer is of high speed, it is far behind human beings in its capacity for remembering the faces of other human beings and discriminating a specific face from a crowd; in other words, the capacity to remember and recognize complex patterns. Furthermore, the intelligence and behavior of human being evolve gradually by learning and training and human beings adapt themselves to changes in the environment. Hence the "neuro-computer", which is based on the neural network (NN) model of human beings was developed and is now one of the important studies in the field of information processing systems [1].

For example, as one imagines "red" when looking at "apple", association recalls from one a pair of patterns. The association process is considered to play the most important role in the intelligent actions of creatures and using the intelligent function of a brain, and is one of the main tasks of study in the history of neuro-computing. Associative memory is, for example, applied to the technology of virtual storage in conventional computer architecture and is also studied in other useful domains [2, 3].

Several models of associative memory (e.g. association, feature map) have been proposed. In cross-coupled attractor-type models such as the Hopfield neural network [4] not only its application but in particular its properties have been studied and the guarantee of convergence, absolute storage capacity, relative storage capacity and other properties have been reported in the literature [5, 6].

cell

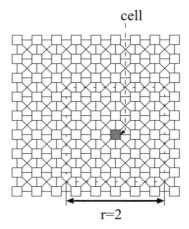

r=2

Fig. 4.1. 9×9 CNN and the $r = 2$ neighborhood

For example, there are some problems in the real world to which syllable recognition by ADALINE, forecasting, noise filtering, pattern classification and the inverted pendulum have been applied. In addition, there are many effective applications: the language concept map using Kohonen's feature map developed with lateral inhibition, facial recognition by concept fuzzy sets by using bidirectional associative memory (BAM) and so forth [7].

On the other hand, cellular automata (CA) are made of massive aggregates of regularly spaced clones called cells, which communicate with each other directly only through nearest neighbors [8, 9]. Each cell is made of a linear capacitor, a non-linear voltage-controlled current source, and a few resistive linear circuit elements. Generally, it is difficult to analyze the phenomena of complex systems. Applying CA to their analysis is expected and being studied.

In 1988, Chua *et al.* proposed the cellular neural network (CNN), which shares the best features of both NN and CA [10, 11, 12]. In other words, its continuous time feature allows real time signal processing, which is lacking in the digital domain and its local interconnection feature makes it tailor made for VLSI implementation. Figure 4.1 shows an example of 9×9 CNN with an r = 2 neighborhood. As shown in Figure. 4.1, the CNN consists of simple analog circuits called cells, in which each cell is connected only to its neighboring cells and it can be described as follows:

$$\dot{x}_{ij} = -x_{ij} + P_{ij} * y_{ij} + S_{ij} * u_{ij} + I_{ij}, \qquad (4.1)$$

where x_{ij} and u_{ij} represent the state and control variables of a cell (i,j) (i-th row, j-th column), respectively, I_{ij} is the threshold and P_{ij} and S_{ij} are the template matrices representing the influences of output or input from the neighborhood cells. The output function y_{ij} is the function of the state x_{ij} and can be expressed as

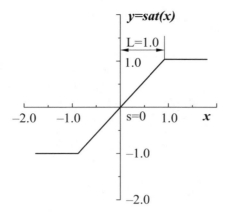

Fig. 4.2. Binary output function

$$y_{ij} = \frac{1}{2}(|x_{ij} + 1| - |x_{ij} - 1|). \tag{4.2}$$

This function is a piecewise linear function as in Figure 4.2, where L is the length of the non-saturation area. Therefore, Equation 4.1 becomes a linear differential equation in the linear regions. When a cell (i, j) is influenced by its neighborhood cells r-units away (see Figure 4.1), $P_{ij} * y_{ij}$ can be expressed as follows:

$$P_{ij} * y_{ij} = \begin{bmatrix} p_{ij(-r,-r)} & \cdots & p_{ij(-r,0)} & \cdots & p_{ij(-r,r)} \\ \vdots & \ddots & \vdots & \ddots & \vdots \\ p_{ij(0,-r)} & \cdots & p_{ij(0,0)} & \cdots & p_{ij(0,r)} \\ \vdots & \ddots & \vdots & \ddots & \vdots \\ p_{ij(r,-r)} & \cdots & p_{ij(r,0)} & \cdots & p_{ij(r,r)} \end{bmatrix} * y_{ij}, \tag{4.3}$$

$$= \sum_{k=-r}^{r} \sum_{l=-r}^{r} p_{ij(k,l)} y_{i+k,j+l}.$$

One can also define $S_{ij} * u_{ij}$ by using the above equation.

As is shown above, the CNN is a nonlinear network, and consists of simple analog circuits called cells. It has been applied to noise removal and feature extraction, and has proved to be effective. In addition, indications show that CNN might be applicable to associative memory.

Liu *et al.* [13] designed CNN for associative memory by making memory patterns correspond to equilibrium points of the dynamics by using a singular value decomposition, and have shown that CNN is effective for associative memory. Since then, its theoretical properties and application to image processing, pattern recognition and so forth have been investigated enthusiastically [14, 15]. CNN for associative memory have the following beneficial properties:

1. The CNN can be represented in terms of a matrix and implemented by a simple *cell* (Figure 4.1), which can be considered an information unit and can be created easily with an electronic circuit.
2. Each cell in the network is connected only to its neighbors. Therefore, its operation efficiency is better than that of all connected neural networks such as Hopfield network (HN).
3. In the case of the classification problem, an improvement of the efficiency and the investigation of the cause incorrect recognition are easy, since the information of the memory patterns are included in the template, which shows the connected state of each cell and its neighbors.
4. In the case of design, adding, removing and modifying memory patterns is easy; like a HN, this does not require constraint conditions such as orthogonally and like a multi-layered perceptron, it does not require trouble some learning.

For these reasons, CNN has attracted attention as an associative memory especially in recent years. Furthermore, in order to improve the efficiency of CNN for associative memory, CNN have been developed into some new models, such as multi-valued output CNN [16, 17, 18] and multi-memory tables CNN [19, 20], and applied to character recognition, the diagnosis of liver diseases, abnormal car sound detection, parts of robot vision systems and so forth [21, 22, 23, 24, 25, 26].

In this chapter, focusing on CNN for associative memory, we first introduce a common design method by using a Singular Value Decomposition (SVD) [27] and discuss its characteristic. We then introduce some new models of the multi-valued output CNN and the multi-memory tables CNN, and their applications to intelligent sensing and diagnosis.

4.2 Design Method of CNN

4.2.1 A Method Using Singular Value Decomposition

As shown in Figure 4.1, CNN consists of simple analog circuits called cells, in which each cell is connected only to its neighboring cells and the state of each cell changes by the differential Equation 4.1.

For simplicity, one assumes the control variable $u_{ij} = 0$ and expresses each cell given in Equation 4.1 as follows:

$$\dot{x} = -x + Ty + I, \tag{4.4}$$

where T is a template matrix composed of row vectors, x is a state vector, y an output vector, and I represents the threshold vector,

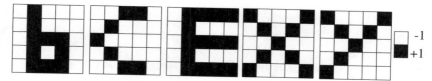

$$-1$$
$$+1$$

Fig. 4.3. Example of memory patterns

$$\left. \begin{array}{l} x = (x_{11}, x_{12}, \ldots, x_{1n}, \ldots, x_{m1}, \ldots, x_{mn})^T \\ y = (y_{11}, y_{12}, \ldots, y_{1n}, \ldots, y_{m1}, \ldots, y_{mn})^T \\ I = (I_{11}, I_{12}, \ldots, I_{1n}, \ldots, I_{m1}, \ldots, I_{mn})^T \end{array} \right\}.$$

$$(4.5)$$

In order to construct CNN, one needs to solve T and I given $\alpha_1, \alpha_2, \ldots, \alpha_q$, which are shown in Figure 4.3 (the pattern with m rows, n columns). These vectors are considered as memory vectors and have elements of $-1, +1$ (the binary output function shown in Figure 4.2). Following Liu and Michel [13], we assume vectors β_i $(i = 1, \ldots, q)$ instead of x at the stable equilibrium points:

$$\beta_i = K\alpha_i, \qquad (4.6)$$

where α_i are the output vectors and K is a location parameter of stable equilibrium points. $K > L$, which shows that K is dependent on the characteristics of the output function $y =$sat(x). Therefore, the CNN that uses $\alpha_1, \alpha_2, \ldots, \alpha_q$ as its memory vectors has a template T and threshold vector I, which satisfies the following equations simultaneously:

$$\left. \begin{array}{l} -\beta_1 + T\alpha_1 + I = 0 \\ -\beta_2 + T\alpha_2 + I = 0 \\ \cdots \\ -\beta_q + T\alpha_q + I = 0 \end{array} \right\}. \qquad (4.7)$$

Let matrices G and Z be

$$\left. \begin{array}{l} G = (\alpha_1 - \alpha_q, \alpha_2 - \alpha_q, \ldots, \alpha_{q-1} - \alpha_q) \\ Z = (\beta_1 - \beta_q, \beta_2 - \beta_q, \ldots, \beta_{q-1} - \beta_q) \end{array} \right\}. \qquad (4.8)$$

In Equation 4.7, we can obtain the following equations by subtracting each equation from the equation by α_q, β_q and by using a matrix expression of Equation 4.8:

$$Z = TG, \qquad (4.9)$$

$$I = \beta_q - T\alpha_q. \qquad (4.10)$$

In order to use α_i as CNN memory vectors, it is necessary and sufficient that the template matrix T and threshold vector I satisfy Equations 4.9 and 4.10. Let us consider the k-th cell in CNN; the conditional equation is given by $(k = n(i - 1) + j)$

$$z_k = t_k G, \tag{4.11}$$

where, z_k and t_k are the k-th row vectors of matrices Z and T, respectively. Using the property of the r neighborhood, we obtain Equation 4.12 by excluding elements that do not belong to the r neighborhood from z_k, t_k and G:

$$z_k^r = t_k^r G^r, \tag{4.12}$$

where G^r is a matrix obtained after removing those elements that do not belong to r neighborhood of the k-th cell from G; similarly, we obtain z_k^r and t_k^r. As a result, we are able to avoid unnecessary computation. The matrix G^r is generally not a square matrix. Therefore, it can be solved by using SVD [27] as follows:

Table 4.1. Each component of vector t_2 when K=1.1

$$\begin{bmatrix} -0.79 & -0.39 & 2.69 & 2.69 & 0.71 \\ 0.0 & 0.0 & 0.0 & -0.79 & 0.0 \\ 1.18 & -0.39 & -1.18 & -0.39 & 1.51 \\ 0.0 & 0.0 & 0.0 & -1.70 & 0.0 \\ -0.79 & -0.39 & 0.79 & 0.79 & 1.70 \end{bmatrix}$$

$$G^r = U_k \cdot [\lambda]^{1/2} V_k^T. \tag{4.13}$$

Hence we have

$$t_k^r = z_k^r V_k [\lambda]^{-1/2} U_k^T. \tag{4.14}$$

This solution is the minimum norm of Equation 4.12, where $[\lambda]^{-1/2}$ is a diagonally dominant matrix consisting of the square root of the eigenvalue of the matrix $[G^r]^T G^r$, and U_k, V_k are the unit orthogonal matrices. In this way, one can construct a CNN whose memory pattern theoretically corresponds to each stable equilibrium point of the differential equation. It is able to associate one pattern by solving Equation 4.4.

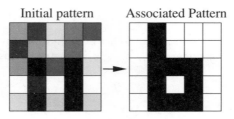

Initial pattern   Associated Pattern

Fig. 4.4. Example of detection results obtained by CNN

Table 4.1 shows the examples of each component of vector t_2 of the template matrix T obtained by using the design method shown above and the pattern shown in Figure 4.3 as memory patterns. When an initial state x_0 (or initial pattern) shown in Figure 4.4 is given, the designed CNN changes each cell's state dynamics by the differential equation Equation 4.4 and converges on the memory pattern shown in Figure 4.4, which is a stable equilibrium point of the optimal solution of the differential equations.

However, the problems in such a CNN are that the influence of the characteristics of the output function and the parameter K have not been taken into account. Therefore, in next Sect., the performance of the output function and the parameter K will be discussed.

4.2.2 Multi-output Function Design

A. Design Method of Multi-valued Output Function

We here show a design method of the multi-valued output function for associative memory CNN. We first introduce the notation that shows how to relate Equation 4.2 to the multi-valued output function. The output function of Equation 4.2 consists of a saturation range and a non-saturation range. We define the structure of the output function such that the length of the non-saturation range is L, the length of the saturation range is cL, and the saturated level is $|y| = H$, which is a positive integer (refer to Figure 4.5). Moreover, we assume the equilibrium points $|x_e| = KH$. Here, the Equation 4.2 can be rewritten as follows:

$$y = \frac{H}{L}(|x + \frac{L}{2}| - |x - \frac{L}{2}|). \qquad (4.15)$$

Then, the equilibrium point arrangement coefficient is expressed as $K = (\frac{L}{2} + cL)/H$ by the above-mentioned definition. When $H = 1, L = 2, c > 0$, Equation 4.15 is equal to Equation 4.2. We will call the waveform of Figure 4.5a the "basic waveform". Next we give the theorem for designing the output function.

Theorem 4.1. *Both $L > 0$ and $c > 0$ are necessary conditions for convergence to an equilibrium point.*

Proof. We consider the cell model Equation 4.1, where $r = 0$, $I = 0$ and $u = 0$. The cell behaves according to the following differential equation:

$$\dot{x} = -x + Ky. \qquad (4.16)$$

In the range of $|x| < \frac{L}{2}$, the output value of a cell is $y = \frac{2H}{L}x$ (refer to Figure 4.5a). Equation 4.16 is expressed by the following:

$$\dot{x} = -x + K\frac{2H}{L}. \qquad (4.17)$$

The solution of the equation is:

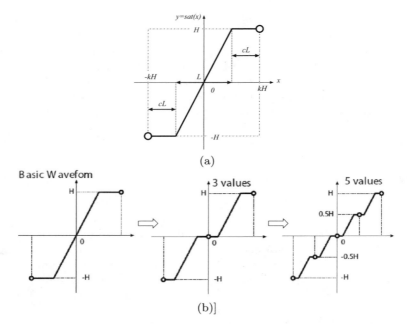

(a)

(b)]

Fig. 4.5. Design procedure of the multi-valued output function: (a) basic waveform and (b) multi-valued output function

$$x(t) = x_0 e^{(\frac{2KH}{L} - 1)t}, \tag{4.18}$$

where x_0 is an initial value at $t = 0$. The exponent in Equation 4.18 must be $\frac{2KH}{L} - 1 > 0$ for transiting from a state in the non-saturation range to a state in the saturation range. Here, by the above mentioned definition, the equilibrium point arrangement coefficient is expressed as:

$$K = (c + \frac{1}{2})\frac{L}{H}. \tag{4.19}$$

Therefore, parameter conditions $c > 0$ can be obtained from Equations 4.18 and 4.19. In the range of $L \le |x| \le KH$, the output value of a cell is $y = \pm H$. Then Equation 4.16 is expressed by the following:

$$\dot{x} = -x \pm KH. \tag{4.20}$$

The solution of the equation is:

$$x(t) = \pm KH + (x_0 \mp KH)e^{-t}. \tag{4.21}$$

When $t \to \infty$, Equation 4.21 proves to be $x_e = \pm kH$, which is not $L = 0$ in Equation 4.19. The following expression is derived from the above:

$$L > 0 \land c > 0. \tag{4.22}$$

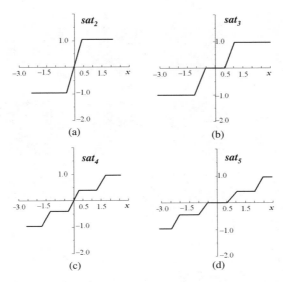

Fig. 4.6. Example of the output waveforms of the saturation function: (a), (b), (c), and (d) represent, respectively, sat$_2$, sat$_3$, sat$_4$ and sat$_5$. Here, the parameters of the multi-valued function are set to $L = 0.5, c = 1.0$

Second, we give the method of constructing the multi-valued output function based on the basic waveform. The saturation ranges with n levels are generated by adding $n - 1$ basic waveforms. Therefore, the n-valued output function sat$_n(\cdot)$ is expressed as follows:

$$\text{sat}_n(x) = \frac{H}{(n-1)L} \sum_i (-1)^i (|x + A_i| - |x - A_i|), \qquad (4.23)$$

$$A_i = \begin{cases} A_{i-1} + 2cL & (i : \text{odd}) \\ A_{i-1} + L & (i : \text{even}) \end{cases}$$

However, i and K are defined as follows:

$$n = \text{odd}: \ i = 0, 1, \ldots, n-2, A_0 = \frac{L}{2}, K = (n-1)(c+1/2)\frac{L}{H},$$
$$n = \text{even}: i = 1, 2, \ldots, n-1, A_1 = cL, K = (n-1)(2c+1)\frac{L}{2H}.$$

Figure 4.6 shows the output waveforms resulting from Equation 4.23. The results demonstrate the validity of the proposed method, because the saturation ranges of the n-levels have been made in the n-value output function: sat$_n(\cdot)$.

B. Computer Experiment

We then show a computer experiment conducted using numerical software in order to demonstrate the effectiveness of the proposed method. For this

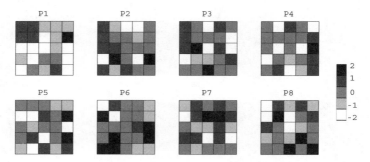

Fig. 4.7. Memory patterns for the computer experiment. These random patterns of five rows and five columns have elements of $\{-2,-1,0,1,2\}$ and are used for creation of the associative memory

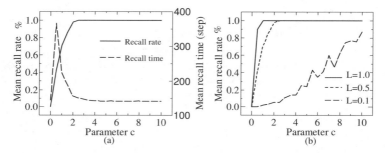

Fig. 4.8. Results of the computer experiments when the standard deviation σ is 1.0: (a) recall rate and time ($L = 0.5$), (b) recall rate and time ($L = 0.1, 0.5, 1.0$)

memory recall experiment, the desired patterns to be memorized are fed into the CNN, which are then associated by the CNN. In this experiment, we use random patterns with five values for generalizing the result as memory patterns. To test recall, noise is added to the patterns shown in Figure 4.7 and the resulting patterns are used as initial patterns. The initial patterns are represented as follows:

$$x_0 = K\alpha^i + \varepsilon, \tag{4.24}$$

where, $\alpha^i \equiv \{x \in \Re^m; x_i = -H, -H/2, 0, H/2, H; i = 1, \ldots, m\}$, and $\varepsilon \in \Re^m$ is a noise vector corresponding to the normal distribution $N(0, \sigma^2)$. These initial patterns are presented to the CNN and the output is evaluated to see whether the memorized patterns can be remembered correctly. Then, the numbers of correct recalls are converted into a recall probability that is used as the CNN performance measure. The parameter L of the output function is in turn set to $L = 0.1, 0.5, 1.0$, and parameter c is changed by 0.5 step sizes in the range of 0 to 10. Moreover, the noise level is a constant $\sigma = 1.0$, and the experiments are repeated for 100 trials at each parameter combination (L, c).

Figure 4.8 shows the results of the experiments. Figure 4.8a shows an example of the relationship between the parameter c and both time and recall probability when $L = 0.5$. Figure 4.8b shows the relationship between the parameter c and recall probability when $L = 0.1, 0.5, 1.0$. The horizontal axis is parameter c and the vertical axes are the mean recall rate (the mean recall probability measured in percent) and mean recall time (measured in time steps).

It is clear from Figure 4.8 that the recall rate increases as parameter c increases for each L. The reason is that c is the parameter that determines the size of a convergence range. Therefore, the mean recall rate improves by increasing c. On the other hand, if the length L of the non-saturated range is short as shown in Figure 4.8b, convergence to the right equilibrium point becomes difficult because the distance between equilibrium points is small. Additionally, as shown in Figure 4.8b, the recall capability is $L = 1.0 > 0.5 > 0.1$. Therefore, the length of the saturation range and the non-saturation range needs to be set at a suitable ratio. Moreover, in order for each cell to converge to the equilibrium points, both $c > 0$ and $L > 0$ must hold.

Therefore, we can conclude that the design method of the multi-valued output function for CNN as an associative memory is successful and we will apply this method for the CNN in an abnormality detection system.

4.2.3 Un-uniform Neighborhood

A. Neighborhood Design Method

In conventional CNN, the neighborhood r is designed equally around any cell. Consequently, the design improving the efficiency has not yet considered the neighborhood. In this Sect., a novel un-uniform neighborhood design method[26] is explained. The neighborhood of each cell is determined in accordance with the following conditions.

1. If a cell has same state for every memory pattern, its neighborhood $r = 0$ will be set because it is not influenced by its neighbor cells.
2. The neighborhood r of the cells that do not conform to condition 1 is determined so that the state of the neighboring cells can differ by at least N cells among the memory patterns. Notice that N should be determined so that the classification capability cannot be lowered.
3. The connection computation of the cells which conform to condition 1 in the range of the neighborhood r of condition 2 is omitted, because their connection coefficients are zero.

Figure 4.9 shows ten model patterns that were used to show the design method by simulations. The design method will be first explained by an example. The cell C(1,2) in Figure 4.9 is an example of a cell that satisfies condition 1. Hence, it has the same state -1 for each memory pattern and is

Fig. 4.9. Model patterns having elements of +1, 0,-1

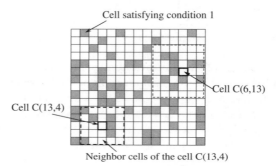

Fig. 4.10. Neighbor cells designed by the new method

not influenced by its neighbor cells. In this case, the differential equation of C(1,2) is expressed as the following:

$$\dot{x}_{1,2} = -x_{1,2} + I_{1,2}. \tag{4.25}$$

From Equation 4.25, one can see that the state of cell C(1,2) will converge to $I_{1,2}$ when an initial state was given, and it is unrelated to its initial state and neighboring cells. Therefore, it is appropriate to set $r = 0$. Similarly, the 60 cells (*e.g*, C(1,14), C(5,1)) shown in Figure 4.10 can be picked up from Figure 4.9.

Next, when $N \geq 15$ in conditions 2 and 3, the example of the neighborhood of the cell C(13,4) is $r = 2$, and its neighbor pattern is shown in Figure 4.10. However, the neighborhood of the cell C(6,13) becomes $r = 3$. That is, a different neighborhood can be set for each cell when N is constant. It is expected that an efficient design of the neighborhood can be achieved by using the method described above and determining the neighbor cells of each cell.

B. Examination Using Model Patters

Here, we first considers to how determine the optimum N. The example of the N of condition 2 is changed in the range of $6 \leq N \leq 30$ by using the model patterns shown in Figure 4.9, and pattern classification is performed, where the CNN size is 15×15. The parameter ρ_0 is 0.38, where ρ_0 is the degree of similarity between the memory patterns and it is a maximum value therein. Here, the degree of similarity ρ indicates the rate of the cell whose state is the same between two patterns. In the case of a 15×15 CNN and $\rho = 0.38$, the number of cells that have the same state is 85 (225×0.38). The initial patterns whose degree of similarity ρ_1 with the memory patterns in 0.8 are used. Ten patterns per memory pattern, that is, 100 patterns are used in all.

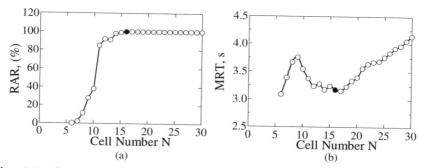

Fig. 4.11. Relation between RCR, MCT and N, where the CNN size is 15×15, $\rho_0 = 0.38$. The calculation was made with a Pentium II 450MHz CPU and Visual $C^{++}6.0$: (a) the relation between N and RAR, (b) the relation between N and MRT

The simulation results are shown in Figure 4.11. Figure 4.11a shows the relation between N and the right cognition rate (RCR,%) and Figure 4.11b shows the relation between N and the mean converging Time (MCT) of 100 initial patterns. It is clear from Figure 4.11a that the RCR first increases as N increases and a 100% RCR can be obtained when $N \geq 16$. The MCT was shown in Figure 4.11b first increases as N increases suddenly, then it decreases and reaches its minimum value around $N = 16$, and then it increases as N increases again. This phenomenon can be explained with the quick speed of the CNN convergence to the equilibrium state of the differential equations. In the case of $N = 6 - 9$, the neighbor r is too small, so the CNN becomes a system which is very hard to converge because the information from the neighborhood is inadequate. In this case, the MCT becomes long as N increases. In the case of $N = 9 - 16$, the information from the neighborhood increases, and the CNN becomes easy to converge. Therefore, the MCT decreases as N increases. In the case of $N = 16 - 30$, because superfluous information is obtained from the neighborhood, the MCT increases as N increases. That is, $N = 16$ is the optimum value since RCR = 100% and MCT also serves as the minimum value, under the condition of $\rho_0 = 0.38$, the CNN's size is 15×15.

We now consider how the value of N is set. Thereupon, the relation between the maximum degree of similarity ρ_0 and N where the RCR becomes

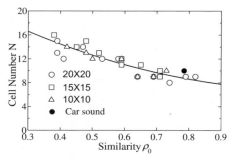

Fig. 4.12. Relation between ρ_0 and N, where CNN are $10 \times 10, 15 \times 15, 20 \times 20$ and the calculating conditions are the same as in Figure 4.11

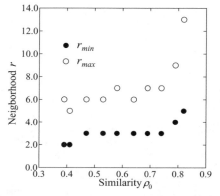

Fig. 4.13. Relation between $r_{\min}, r_{\max}$ and ρ_0, where CNN is 20×20

100%, changing memory patterns was examined. The result is shown in Figure 4.12. Moreover, three kinds of CNN: $10 \times 10, 15 \times 15, 20 \times 20$ used where in order to also examine the relation between N and the CNN size. The initial patterns where the degree of similarity ρ_1 with memory patterns is 0.8 are used. Furthermore, an example of the relation between the neighborhood $r_{\min}$, $r_{\max}$ and the ρ_0 is shown in Figure 4.13, where the $r_{\min}, r_{\max}$ were determined from the same N in Figure 4.12, and the CNN size is 20×20.

As shown in Figure 4.12, N decreases as ρ_0 increases, and N is almost not influenced by the CNN size. In the case of the large ρ_0, $r_{\min}$ and $r_{\max}$ shown in Figure 4.13 have large values, that is, neighbor cells increase as ρ_0 increases even if N is small. The curve in Figure 4.12 shows the second-order approximated. The approximated expression is represented as

$$y = 16.995x^2 - 34.919x + 25.688. \qquad (4.26)$$

Furthermore, the frequency distributions of the neighbor r are shown in Figure 4.14, where $\rho_0 = 0.53$, $\rho_0 = 0.69$ and $\rho_0 = 0.82$. As shown in Fig-

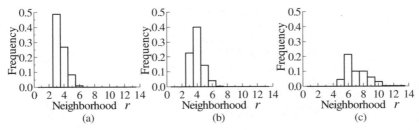

Fig. 4.14. Frequency distribution of the neighbor r, where CNN is 20×20: (a) in the case of $\rho_0 = 0.53$, (b) in the case of $\rho_0 = 0.69$ and (c) in the case of $\rho_0 = 0.82$

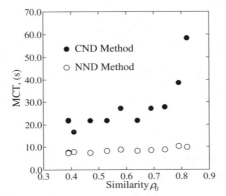

Fig. 4.15. Relation between MCT of the CND method, the NND method and ρ_0, where CNN is 20×20 and calculations were made under the same conditions as Figure 4.11

ure 4.14, the pick of the frequency distributions of r becomes short, the width becomes wide and the centers value of the frequency distributions shift onto large values as ρ_0 becomes large.

In the conventional neighbor design method (CND) of the CNN, the neighborhood r is designed equally about every cell. Therefore, in order to obtain a 100% recognition rate, an $r \geq r_{max}$ value is generally used. On the other hand, in the new neighbor design method (NND), a different neighborhood r shown in Figure 4.10 for each cell is obtained by using fixed N, and the calculation time amount can be reduced and the efficiency of the CNN can be improved. Moreover, the frequency of the cell meeting condition 1 is shown in Figure 4.15. The number of cells adhering to condition 1 increases as ρ_0 increases. It turns out that the amount of calculation of the cells meeting condition 1 is also reduced by our approach. The relation between the maximum degree of similarity ρ_0 and the MCT is shown in Figure 4.15, where the white circles correspond to the NND method and the black dots to the CND method. As shown in Figure 4.15, in the case of the CND, the MCT increases greatly in comparison, although in the case of the NND, the MCT increases slightly as ρ_0 increases. About the average MCT in the range of $\rho_0 = 0.4 - 0.7$, NND is 8.07 s and CND is 23.09 s, and MCT of NND is 35% of CND can be obtained. Moreover, about the average MCT around $\rho_0 = 0.8$, NND is 10.04 s and CND is 48.53 s, and MCT of CND is 21% of CND, that is, the improvement rate becomes high as the ρ_0 becomes large.

Therefore, if one designs the CNN so that the value of N becomes somewhat larger than the approximated curve shown in Equation 4.26, then maintaining a high classification capability and a reduction in the recall time can be achieved.

4.2.4 Multi-memory Tables for CNN

If CNN can memorize and classify many and similar patterns, then they can be applied to other fields. However, they do not always work well.

It is well-known that in CNN, memory patterns correspond to the equilibrium points of systems, and the state of each cell changes by the influences of neighbor cells. Finally, networks converge on them. That is, CNN have a self-recall function. However, in the dynamics of CNN, the network does not always converge on them when embedding patterns too many and including similar patterns. These cases called "incomplete recall". Fortunately, the most appropriate pattern number or its range that maximizes the self-recall capability exists in each CNN for associative memory. Based on this, a new model of the CNN with multiple memory tables (MMT-CNN), in which multiple memory tables are created by divisions of a memory pattern group, and the final recall result is determined based on the recall results of each memory table, was considered [19, 20]. In this Sect., we will introduce the basic theory of MMT-CNN and discuss its characteristics.

In order to design MMT-CNN, the capability of conventional CNN(the relations between the number of memory patterns, their similarities and the self-recall capability) is first confirmed. To this end, the similarity of patterns should be defined quantitatively. In this Sect., the Hamming distance d is used first. Hamming distance between two vectors $a = (a_1, a_2, \ldots, a_N)$, $b = (b_1, b_2, \ldots, b_N)$ is defined as follows:

$$d = \sum_{k=1}^{N} \delta(a_k, b_k), \qquad (4.27)$$

where

$$\delta(a_k, b_k) = \begin{cases} 0 : a_k = b_k \\ 1 : a_k \neq b_k \end{cases} \qquad (4.28)$$

Then the minimum d_0 in the Hamming distances of any two memory patterns is the distance of the memory pattern group. In addition, the following parameter D_0, which does not depend on CNN size, is defined.

$$D_0 = \frac{d_0}{N_0}, \qquad (4.29)$$

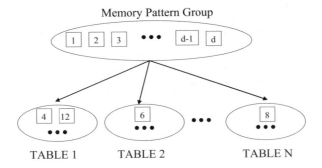

Fig. 4.16. The method of dividing memory patterns

where N_0 denotes the cell number. Hence, when D_0 is small, its pattern group has a high similarity.

It is well-known that CNN have a local connectivity. However, more information from neighbor cells improve the recall ability of CNNs. Hence, in order to avoid the influence of neighbor size and to obtain the maximal associated ability of CNN, the full connected CNN are used in the computational experiments. Furthermore, the binary random patterns are used as memory patterns so as to keep the generality. The initial patterns by multiplied Gaussian noise are given CNN.

In the condition shown above, the maximum of capacity so that the incomplete recall rate (ICR) can be 0%, changing the number of memory patterns have been considered. Where the initial pattern generated by multiplying the memory pattern by Gaussian noise and five kinds of Gaussian noise with different strength σ have been used. Furthermore, 6000 initial patterns are generated per σ and they have been used to recall experiments. The average of their values is called "limit memory patterns". It is denoted by $M_{\lim}(m, n)$ in $m \times n$ CNN. The relation between the strength of Gaussian noise $\sigma = 1.0$ and $M_{\lim}(m, n)$ is examined and the obtained relation between $M_{\lim}$ and N_0 in $\sigma = 1.0$ can be approximated as follows:

$$M_{\lim}(m, n) = 0.45N_0 - 10. \tag{4.30}$$

From Equation 4.30, it is clear that full connected CNN can memorize patterns of about 45% (the second term in Equation 4.30 can be ignored when N is sufficiently large). Therefore, Equation 4.30 is applied to the design of MMT-CNN.

The procedure of MMT-CNN consists of two steps: 1) classification and divisions of memory patterns shown in Figure 4.16, and 2) the associated algorithm shown in Figure 4.17. In the first step, the following two conditions are used.

1. $D_0 \leq 0.05$
   According to [14], when the degree of similarity ρ between input and

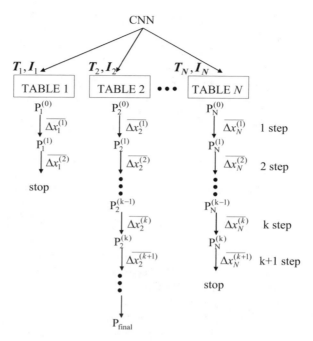

Fig. 4.17. Algorithm of MMT-CNN

desired patterns is more than 80%, CNNs can generally recall correctly. We can obtain the above condition by replacing the similarity by the Hamming distance Equation 4.29.

2. $M \geq M_{\lim}(m, n)$

The number of memory patterns should be restricted in order to maintain the reliability of CNNs. Hence, the condition as the described above is set.

If any of the above conditions are satisfied, then all memory patterns have to be divided into N memory tables in MMT-CNN in order for D_0 to enlarge. Of course, if the conditions have not satisfied, then the common CNN should be used. Furthermore, in order to reduce the load of divisions the simple division algorithm shown in Figure 4.16 is used.

1. Select a pattern at random in memory patterns, and set it in TABLE 1.
2. Find the distances between the patterns selected in 1 and the remains, and set the pattern that gives the least distance in TABLE 2.
3. Repeat 2 about the pattern selected in 2 and remainder.
4. Find the distances among constructed TABLEs. If they do not satisfy $D_0 \leq 0.05$, then start over after changing the division number and replacing patterns.

After setting the TABLEs, following Sect. 4.2.1, the template matrix T and threshold vector I about each memory TABLE are designed. Then the behavior of MMT-CNNs (see Figure 4.17) can be shown as follows:

1. Find the template matrix T and threshold vector I about each memory TAVBLE.
2. Consider TABLE 1 as memory patterns, and recall by 1 step from the initial pattern.
3. Considering TABLE 2 as memory patterns, and recall by 1 step from the initial pattern. The procedures are iterated by TABLE N. These procedures are considered as one step of MMT-CNN. If CNN converge in either TABLE, the calculation is finished.
4. For efficiency, the mean of state varies $\overline{\Delta x_i}(i = 1, \ldots, N)$ in every TABLE in 3 are found. Let the maximum and the minimum be $\overline{\Delta x_{\max}}$, $\overline{\Delta x_{\min}}$ respectively. When Equation 4.31 is satisfied, the TABLE giving $\overline{\Delta x_{\max}}$ can be removed,

$$\overline{\Delta x_{\max}} - \overline{\Delta x_{\min}} > c\overline{\Delta x_{\max}}, \tag{4.31}$$

where the constant c satisfies $0 < c < 1$; $c = 0.3$ was set for accuracy and efficiency.
5. Returning to 1. and continuing the recall procedures.

The state of CNN changes by the dynamics of differential equations. Hence, if the amount of changes is small, then the CNN close to the convergence can be considered. Consequently, step 4 gives faster processing.

By the new model of the MMT-CNN, the network size can merely be enlarged in order to increase the memory capacity. However, the similarity of memory patterns in this method cannot be reduced. Consequently, the model of the MMT-CNN is superior to the method described above. In order to show the performance of the MMT-CNN, its application in the pattern classification will be shown in Sect. 4.3.3.

4.3 Applications in Intelligent Sensing and Diagnosis

4.3.1 Liver Disease Diagnosis

Most inspection indices of blood tests consist of three levels, for example, γ–GTP has roughly the following three levels: normal (0–50 IU/l), light excess (50–100 IU/l), and heavy excess (100–200 IU/l). Also, for example, ChE 200–400 IU/l is considered as the normal value, but the possibility of hepatitis or the fat liver is diagnosed when it is lower or higher than the normal value, respectively. In this Sect., we apply the CNN of $r = 4$ based on the tri-valued output function in Equation 4.23 to classify liver illness [17, 18]. Following Figure 4.8, the parameters of the tri-valued output function were set as $H =$

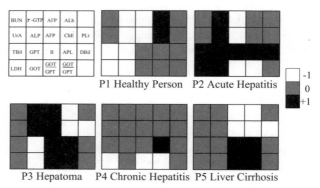

BUN	γ-GTP	AFP	ALb	
UrA	ALP	AFP	ChE	PLt
TBil	GPT	II	APL	DBil
LDH	GOT	GOT/GPT	GOT/GPT	

P1 Healthy Person   P2 Acute Hepatitis

☐ -1
▓ 0
■ +1

P3 Hepatoma   P4 Chronic Hepatitis P5 Liver Cirrhosis

Fig. 4.18. Pattern of five liver illness

1.0, $L = 0.5$, $c = 2.0$ and the shape of the tri-valued output function is shown in Figure 4.6b.

The blood test data provided by the Kawasaki Medical School [28] were collected from patients suffering from liver diseases. The data set represents five liver conditions: healthy (P1), acute hepatitis (P2), hepatoma (P3), chronic Hepatitis (P4) and liver cirrhosis (P5). Moreover, 50 patients' data for every illness (a total of 250 people), and 20 items of blood test results, such as γ–GTP for each patient are given. The data set has large variations, with missing items and spurious values due to instrumental imperfection. Since these samples are good representatives of the problems present in most clinical diagnostic processes, we can evaluate the performance of our proposed method using this data set to verify its usefulness in practice.

We here first distribute each standard value of blood based on medical specialists into three levels of -1, 0, $+1$, which is shown in Table 4.2 and use the five liver disease patterns, which are stored in the 4×5 CNN matrix shown in Figure 4.18 as the memory patterns. Following the steps detailed in Sects. 4.2.1 and 4.2.2, we then constructed CNN using parameter K obtained by Equation 4.19, which corresponds to the tri-valued output function Equation 4.23 described in the previous Sect. for all cells and classified the 250 patients' data.

Table 4.3 shows the diagnostic results recalled by the CNN, where, row P3 and column P4 in Table 4.3 indicate a patient who should belong to P3 but was classified as P4, which represents a misdiagnosis. The values in the "other" column are the patient numbers that could not be diagnosed because the associated patterns did not belong to any memory pattern. Table 4.3 shows that using the tri-valued output function shown in Equation 4.23 we were able to obtain a 100% correct diagnosis rate (CDR) for healthy persons and acute hepatitis, whereas for Hepatoma, chronic hepatitis and liver cirrhosis, we were able to obtain, on average a 70% CDR .

As a comparison, Yanai [29] reported the results of diagnosis by if-then rules using the rough set theory and fuzzy logic, where a part of data, that

Table 4.2. Scaling function by consulting medical standards

q_k: Parameters	d_k: Blood tests	Medical names
$q_1 = (d_1 - 15.0)/12.0$	d_1: BUN	Blood Urea Nitrogen
$q_2 = d_2/50.0 - 1.0$	d_2: γ-GTP	γ-Glutamyl Transpeptidese
$q_3 = \log(d_3/50.0)$	d_3: AFP	Alpha-1 Fetoprotein
$q_4 = (d_4 - 3.95)/0.75$	d_4: Alb	Albumin
$q_5 = d_5$	$d_5 = 0.0$:	Nothing
$q_6 = (d_6 - 5.0)/5.0$	d_6: UrA	Uric Acid
$q_7 = (d_7 - 90.0)/60.0$	d_7: ALP	Alkaline Phosphates
$q_8 = q_3$	d_3: AFP	Alpha-1 Fetoprotein
$q_9 = (d_9 - 225.0)/125.0$	d_9: ChE	Cholinesterase
$q_{10} = (d_{10} - 25.0)/20.0$	d_{10}: PLt	Plate Let
$q_{11} = \log(d_{11}/50.0)/0.5$	d_{11}: Tbil	Total Bilirubin
$q_{12} = \log(d_{12}/90.0)$	d_{12}: GPT	Glutamic Pyruvic Transaminase
$q_{13} = (d_{13} - 5.0)/3.8$	d_{13}: II	Ic-Terus
$q_{14} = 0.0$ (Lacking data)	d_{14}: LAP	Leucine Aminopeptidase
$q_{15} = (d_{15} - 45.0)/30.0$	d_{15}: Dbil	Direct Bilirubin
$q_{16} = (d_{16} - 175.0)/120.0$	d_{16}: LDH	Lactate Dehydrogenase
$q_{17} = \log(d_{17}/90.0)$	d_{17}: GOT	Glutamic Oxaloacetic Transaminase
$q_{18} = q_{17}/q_{12}$	d_{18}: GOT/GPT	Ratio of GOT to GPT
$q_{19} = q_{17}/q_{12}$	d_{19}: GOT/GPT	Ratio of GOT to GPT
$q_{20} = d_{20}$	$d_{20} = 0.0$	Nothing

Table 4.3. Diagnosis results recalled by the CNN

Liver diseases	No.	P1	P2	P3	P4	P5	Other	CDR %
P1 Health Person	50	50					0	100%
P2 Acute Hepatitis	50		50				0	100%
P3 Hepatoma	50			35	1	1	13	70%
P4 Chronic Hepatitis	50			3	40	2	5	80%
P5 Liver Cirrhosis	50		1		4	30	15	60%

is, 20 patients' data for every illness (a total of 100 people) were used. They were able to achieve a good CDR: healthy person is 100%, the acute hepatitis 72%, hepatoma 59%, chronic hepatitis 62% and liver cirrhosis 65%. Due to the lack of details of their data, we are not able to carry out a direct comparison. Nevertheless, we can see that our system has indeed performed at least as well as that reported in [29].

Furthermore, a comparison of the CNN with that obtained by the more conventional three layer, feed forward neural network (NN) shown in Figure 4.19a was made. The input layer of the NN had 20 units corresponding to the amount of the features shown in Figure 4.19. The number of hidden layer units was 40 as a result of trials, and the number of output layer units was five for five types of liver disease. In the case of the input layer, a pair of units

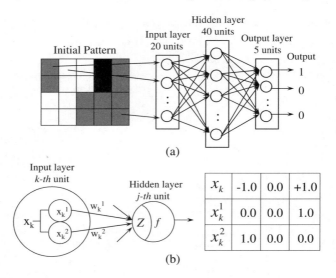

(a)

(b)

Fig. 4.19. Structure of the NN and a pair of input units: (a) shows structure of the NN and (b) each input unit has a pair of units

Table 4.4. Diagnosis results recalled by Perceptron type neural network

Liver diseases	No.	P1	P2	P3	P4	P5	Other	CDR %
P1 Healthy Person	50	50					0	100%
P2 Acute Hepatitis	50		46				4	92%
P3 Hepatoma	50			31	2	3	13	62%
P4 Chronic Hepatitis	50	1	2		35	1	11	70%
P5 Liver Cirrhosis	50	1	1	2	6	26	14	52%

was used as each input unit shown in Figure 4.19b in order to achieve the tri-valued input for the NN. For example, when the state of the k-th item in a initial pattern $x_k=-1.0$, its corresponding pair of units x_k^1, x_k^2 became $x_k^1=0.0$, $x_k^2=1.0$, and when $x_i=+1.0$, the pair became $x_k^1=1.0, x_k^2=0.0$, etc. Learning was carried out by using the well-known back-propagation algorithm. The data of the typical patterns shown in Figure 4.18 was used as training data. In addition, the learning was repeated until the average square error between the training data and the output value was below 0.005. When the output value was the same as or larger than 0.8, the disease name corresponding to the unit was given as the diagnostic result. If the output value was lower than 0.8, it was considered as an uncertain diagnosis.

Table 4.4 shows the results obtained by the NN. As shown in the table, the average CDR of a healthy person and acute hepatitis was 96%, and average CDR of Hepatoma, chronic hepatitis and liver cirrhosis was 61%. This indeed has clearly shows that the CNN performed better than that of the conventional NN.

4.3.2 Abnormal Car Sound Detection

CNN is also applied to diagnose abnormal automobile sounds [23, 26]. The abnormal sounds are contained in the mimic sound expression for vehicle Noise, which sold by the Society of Automotive Engineering of Japan [30]. The measuring conditions of these sounds are various and includes a lot of noise by each part of the car except for abnormal sounds. Each abnormal sound is determined in advance, which is useful for the testing of our proposed method. We chose 12 such kinds of sound signals, and extracted 15 samples from each kind of signal (a total of 169 samples).

A. Maximum Entropy Method (MEM)

In order to extract the characteristics of the signal, the method called the maximum entropy method (MEM), which is a frequency analysis method, was first used [31].

Generally, the AR model for a steady signal is given by:

$$x_n = \sum_{k=1}^{p} a_k x_{n-k} + e_n,$$

(4.32)

where subscript n denotes time which corresponds to $t = n\Delta\tau$, with $\Delta\tau$ the sampling interval, and a_k denotes the coefficient of the AR model, which changes with k. In Equation 4.32, we assume $\sum a_k x_{n-k}$ to be the predicted value of the signal x_n , and e_n to be the prediction error of the signal x_n . The power spectrum of the signal can also be obtained from the AR model, which can be shown as the following formula:

$$E(f) = \frac{2\sigma^2 \Delta\tau}{|1 - \sum_{k=1}^{p} a_k e^{-i2\pi f k \Delta\tau}|^2},$$

(4.33)

where σ^2 denotes the variance of the prediction error and the number of the coefficients in Equation 4.33 is p. In order to calculate coefficient a_k , one can use the Burg algorithm [31], which has the advantage of high resolution for short data under the condition of maximum entropy (so it was called MEM). Moreover, the variable p of the coefficient in Equation 4.33 is an important parameter that influences the stability and resolution of the signal's power spectrum. However, there is no rational standard to determine it. Ulrych *et al.* [31] proposed the following equation and determined p when the final prediction error (FPE) standard attains its minimum in the following equation

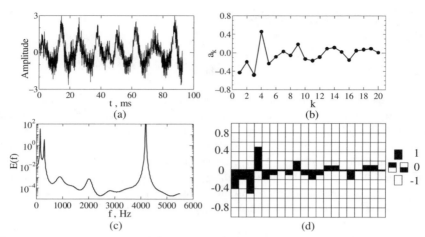

Fig. 4.20. Example of a sound signal and characteristic pattern of a_k: (a) example of sound signal, (b) power spectrum of the sound signal, (c) coefficient a_k by MEM, and (d) characteristic pattern of a_k

$$FPE = \frac{[N + (p + 1)]}{[N - (p + 1)]} \sigma^2, \qquad (4.34)$$

where N is the number of data points ($N\Delta\tau = \Delta t$, Δt denotes data length). Moreover, there exists an Akaike information criterion (AIC) standard, which is also widely used like FPE. In the case of the AR model shown in Equation 4.32, both FPE and AIC are equivalent. Therefore, in this Sect., the FPE standard was used. However, in the case where the signal has a sharp spectrum, the FPE does not converge clearly to a minimum value, so we needed to cut off p in the lower half of the data and the optimum p was determined as in the following equation:

$$p < (2 \sim 3)\sqrt{N}. \qquad (4.35)$$

However, in the case of the car sounds, FPE converges slowly to the minimum value as the number of coefficients increases, not depending on sound signals. However, in order to obtain the minimum of FPE, a sufficiently large p should be chosen. Therefore, following Equation 4.35, we chose $p = 20$ (constant), since FPE is relatively small and it is not changed dramatically.

B. Constitution of the Characteristic Pattern

We here extract the characteristics of the sound signal by the coefficient a_k of the number p from a long sound signal (the number of data is N) using MEM, and make the pattern for CNN using the coefficient a_k of the number p. Then we perform an ambiguous classification of the pattern obtained using CNN.

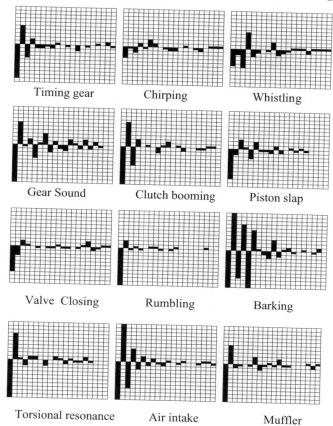

Fig. 4.21. Characteristic patterns of abnormal automobile sounds

Figure 4.20a shows an analysis example of a muffler whistling sound "PEE-" from a car, where the sampling frequency is 11 kHz, and the number of analysis data is 1024. Figure 4.20b shows the power spectrum obtained by MEM, and Figure 4.20c shows the coefficient a_k of auto-regression (AR) model obtained by the MEM representing the characteristic of the signal. As shown in the figures, the power spectrum of sound "PEE-" has a large peak at about 200 Hz and 4 kHz, and its characteristic is represented by only 20 coefficients a_k. That is, the information of the sound signal whose number of data is 1024 is compressed into 20 coefficients a_k, by making its information entropy maximum, and the characteristic of the signal is extracted.

Next, 20 a_k coefficients obtained by MEM are scale-transformed, and the characteristic pattern applying to CNN is constituted. Figure 4.20d shows that the characteristic pattern of the CNN consisted of 10×20 cells, which are scale-transformed from the coefficients shown in Figure 4.20c. One then explain the allocation of the CNN cells. The number of horizontal cells is

Initial pattern

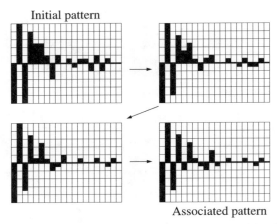

Associated pattern

Fig. 4.22. Example of detection results obtained by CNN

equal to the number of the coefficients a_k and the vertical axis represents the amplitude of the coefficients a_k by combining ten cells. The upper five cells correspond to positive amplitudes and the lower five correspond to negative amplitudes. Furthermore, the amplitude of a_k can be shown by the state of the cell corresponding to black ($+1$), half of the cell is black and the other half is white (0), or white (-1). Consequently, each amplitude of coefficients is represented by the height of black cells shaped like a bar consisting of the state of cells ($+1$ or 0) in the vertical direction.

Then, we take a_4, which is shown in Fig 4.20c as an example, and show the method of scale-transforming coefficients into a characteristic pattern. First, in the approximation of the value in scale-transformation, two decimal places are counted as one fraction of more than 0.5 inclusive. For example, though the value of a_4 is 0.459, it is treated as 0.5 by approximation. Next, each cell of the vertical direction has a scale of 0.2, and the state of the cell changes to black or white by corresponding to the amplitude of the coefficient a_4. The location of $a_4 = 0.5$ is at the upper five cells of the fourth column cells in Figure 4.21 since it is positive, and the state of the first and second cells from the top are white (-1), the third one is half black and half white (0), and the fourth and fifth cells are black ($+1$). Furthermore, the lower cells of the same column are transformed into white (-1). Thus, every coefficient is transformed into the state of each column cell corresponding to itself, and the characteristic pattern of the CNN is obtained.

C. Diagnosing Abnormal Sound by CNN

Figure 4.21 shows the 12 memory patterns obtained by scale-transformation using the method shown above. As shown in Figure 4.21, each memory pattern has each characteristic. These patterns are memorized in CNN, and then 169 sample data of 12 kinds of abnormal signals are input into the CNN as initial

Table 4.5. Discrimination results

Sound Data	No. of data	Right	Error	Other	Right ratio
Sample 1	15	15	0	0	100%
Sample 2	15	15	0	0	100%
Sample 3	15	15	0	0	100%
Sample 4	15	15	0	0	100%
Sample 5	11	11	0	0	100%
Sample 6	15	15	0	0	100%
Sample 7	15	15	0	0	100%
Sample 8	15	15	0	0	100%
Sample 9	15	15	0	0	100%
Sample 10	12	11	0	1	92%
Sample 11	15	15	0	0	100%
Sample 12	11	11	0	0	100%
total	169	168	0	1	99%

patterns and fuzzy discrimination is carried out by CNN. Figure 4.22 shows an example of sound "PEE-" (gear sound) detection process. As is shown in Figure 4.22, the sound "PEE-" has been detected correctly, that is, if the common feature exists, CNN can classify it correctly although the initial pattern differs from the memory pattern.

Table 4.5 shows the discrimination results of 169 sample data. As shown in Table 4.5, the CNN has a high discrimination capability (rate: 99%), synthetically, although an "other" sample is found in 12 initial patterns of the tenth sound "GAH", which means the CNN does not converge on any pattern. By comparing the pattern "other" with the desired pattern (tenth pattern "GAH" in Figure 4.21, we can see that the pattern "other" is close to the desired pattern "GAH". Consequently, it is expected that the discrimination capability can improve by introducing a distance discrimination method.

Comparing it with abnormal sound, the power spectrum of normal white noise is smooth and its coefficients a_k are almost the same. When the coefficients a_k of normal white noise are transformed to the pattern and are input into the CNN as initial pattern, the detection result of "other" is always obtained, that is, the abnormal sound flag does not trigger.

Furthermore, when the design method of un-uniform neighborhood shown in Sect. 4.2.3 was used to design the CNN for the diagnosis of abnormal sounds, the results obtained can be shown as follows: the MCT of the NND is 2.048 s and the MCT of CND is 53.74 s under the conditions of a CPU Pentium II 450 MHz and Visual $C^{++}6.0$. The relative computation time (RCT) is shown in Figure 4.23. The RCT of the NND is expressed relative to the CND where the computation time is set at 100, and, as shown in Figure 4.23, the computation time of NND is only 4.48% of the CND, that is, the un-uniform neighborhood design method shown in Sect. 4.2.3 is indeed effect of improving the CNN's capability.

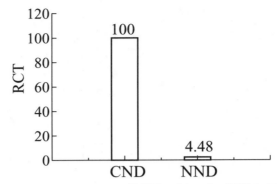

Fig. 4.23. Relative computation time (RCT), where the CNN is 20×20 and $N=10$, which was determined by Figure 4.12

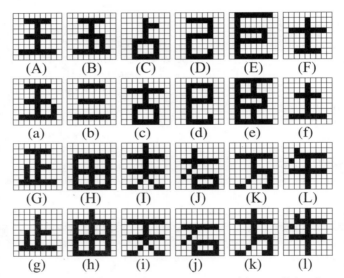

Fig. 4.24. Memory pattern group used in condition 1

4.3.3 Pattern Classification

The 24 Chinese character patterns shown in Figure 4.24 and the 600 figure patterns shown in Figure 4.25 are used in pattern classification experiments, because not only numerous but similar patterns are included in them. In order to show the effectiveness of the MMT-CNN shown in Sect. 4.2.4, the self-recall results of two cases (embedding numerous patterns and including similar patterns) are considered.

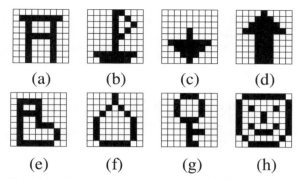

$$(a) \qquad (b) \qquad (c) \qquad (d)$$

$$(e) \qquad (f) \qquad (g) \qquad (h)$$

Fig. 4.25. Memory pattern group used in condition 1

From Sect. 4.2.4, the conditions that should be verify are as follows:
1. $D_0 \leq 0.05, M < M'_{\lim}(m, n)$;
2. $D_0 \leq 0.05, M \geq M'_{\lim}(m, n)$;
3. $D_0 > 0.05, M \geq M'_{\lim}(m, n)$.

Condition 3 is obvious if the conditions 1 and 2 can be achieved. Hence, we examined the conditions of 1 and 2.

First in condition 1, we used a group (1 TABLE) as shown in Figure 4.24, whose distance D_0 is 0.02. The group consists of 24 pairs of extremely similar patterns. On the other hand, the 2, 3 and 4 TABLEs were created by using the division algorithm described in Sect. 4.2.4, and shown in Table 4.6, where the number of patterns per TABLE is 12, 8 and 6, and minimum D_0 is 0.02, 0.05 and 0.09, respectively. As is shown in Table 4.6, each D_0 is decreased by increasing TABLEs, that is, it is recognized that the similarity of memory pattern is decreased.

Figure 4.26a shows the incomplete recall rate (ICR) of the conventional CNN and MMT-CNN, where the vertical axis shows the ICR and the horizontal axis shows the noise level σ of initial patterns. The initial pattern generated by multiplying the memory pattern by Gaussian noise and five kinds of Gaussian noise with different strength σ has been used. Furthermore, 6000 initial patterns are generated at each σ and they have been used to recall experiments, respectively. Finally the average ICRs are obtained. As is shown in Figure 4.26a, in the case of conventional CNNs, the ICR increases as σ increases when $\sigma > 0.4$. Compared to conventional CNN, in the case of two divisions, it can be restrained sufficiently, in the case of three divisions, the ICR is approximately 0% and in the case of four divisions, the ICR is 0%. It can be recognized that in the case of four divisions, all the TABLEs satisfy $D_0 \geq 0.05$. Consequently, it is recognized that the condition of D_0 is reasonable. At the same time, even if the patterns in the memory pattern group are few, MMT-CNN is considered to be useful.

Next in condition 2, the figure patterns $M = 600, D_0 = 0.02$ are used as memory pattern group and the examples are shown in Figure 4.25. In this

Table 4.6. Division results and each D_0 in condition 1: (a) two divisions, (b) three divisions, and (c) four divisions

(a)

Pattern						D_0
(C) (k) (i)	(l)	(H) (e)				
(F) (g) (A)	(B)	(D) (J)				$D_0 = 0.02$
(c) (K) (I)	(L)	(h) (E)				
(f) (G) (a)	(b)	(d) (j)				$D_0 = 0.05$

(b)

Pattern				D_0
(C) (k)	(l)	(h)		
(F) (G)	(B)	(d)		$D_0 = 0.05$
(c) (i)	(L)	(e)		
(f) (A)	(b)	(J)		$D_0 = 0.05$
(k) (I)	(H)	(E)		
(g) (a)	(D)	(j)		$D_0 = 0.11$

(c)

Pattern						D_0
(C) (i) (H)	(F)	(A) (D)				$D_0 = 0.17$
(c) (I) (h)	(f)	(a) (d)				$D_0 = 0.17$
(k) (l) (e)	(g)	(B) (J)				$D_0 = 0.11$
(K) (L) (E)	(G)	(b) (j)				$D_0 = 0.09$

case, the four kinds of MMT-CNN (10, 15, 20, 25 TABLEs) have been used. Hence, the number of patterns per TABLE is 60, 40, 30, and 24, respectively. As in condition 1, the initial pattern generated by multiplying the memory pattern by Gaussian noise and five kinds of Gaussian noise with different strength σ has been used. Furthermore, 6000 initial patterns are generated at each σ and they have been used to recall experiments, respectively. Finally the average results can be obtained by averaging recall results ICRs of 6000 initial patterns.

Figure 4.26b shows the average ICR of the conventional CNN and MMT-CNN. As is shown in the figure, in the case of the conventional CNNs, ICR is about 100% in the range of $\sigma \geq 0.6$. On the other hand, in MMT-CNN, it sufficiently decreases. When the division number increases, it additionally decreases. Especially, in the case of 20 divisions, the ICRs are 0% approximately and in the case of 25 divisions, the ICRs are 0%. These results approximately correspond with the estimation $M \leq M_{\lim}(m, n)$ ($M \leq 26$ when CNN size is 9×9) shown in Sect. 4.2.4. Furthermore, the effectiveness of MMT-CNN has been confirmed in the conditions of other sized CNN (12×12, 15×15) and almost same results have been obtained. Based on the above discussion, the new model of the MMT-CNN is effective for pattern classification even though

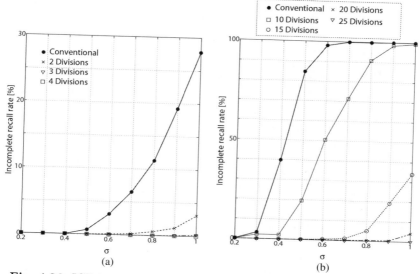

Fig. 4.26. ICR in conditions 1 and 2: (a) condition 1 and (b) condition 2

memory patterns are not only numerous but similar patterns are included in therein.

4.4 Chapter Summary

Recently, various models of associative memory have been proposed and studied. Their tasks are mainly expansion of storage capacity, accurate discrimination of similar patterns, and reduction of computation time. Chua *et al.* [10] described in their paper that CNN can be exploited in the design of associative memories, error correcting codes and fault tolerant systems. Thereafter, Liu *et al.* [13] proposed the concrete design method of CNN for associative memory. Ever since, some applications have been proposed; however, studies on improving its capability are few. Some researchers have already shown CNN to be effective for image processing. Hence, if advanced association CNN system is established, for example, an CNN recognition system can be constituted. Moreover, it will be capable of widely applications.

In this chapter, we focused on CNN for associative memory and first introduced a common design method by using a singular value decomposition and discussed its characteristics. Then we introduced some new models, such as the multi-valued output CNN and the multi-memory tables CNN, and their applications in intelligent sensing and diagnosis. The results in this chapter can contribute to improving the capability of CNN for associative memory.

Moreover, they would indicate the future possibility of CNN as the medium of associative memory.

References

1. Dayhoff J (1996) Neural network architectures : An introduction. International Thomson Computer Press, Boston
2. Kung SY (1993) Digital neural networks. PTR Prentice Hall
3. Haykin S (1994) Neural networks - a comprehensive foundation. Macmillan College Publishing
4. Hopfield JJ (1982) Neural networks and physical systems with emergent collective computational abilities. Proc. of the National Academy Sciences 79:pp.2554–2558
5. Nakano K (1972) Association - a model of associative memory. IEEE Transaction, SMC–2:380–388
6. Kohonen T (1993) Self-organizing map. Proc. IEEE : Special Issue on Neural Networks I, 78:1464–1480
7. Kosko B (1995) Bidirectional associative memory. IEEE Transaction SMC. 18:49–60
8. Wolfram S (ed.) (1986) Theory and applications of cellular automata. New York, World Scientific, Singapore
9. Packard N and Wolfram S (1985) Two-dimensional cellular automata. Journal of Statistical Physics, 38:901–946
10. Chua LO and Yang L (1988) Cellular neural networks: theory. IEEE Transactions on Circuits and Systems, CAS–3:1257–1272
11. Chua LO and Yang L (1988) Cellular neural networks: applications. IEEE Transactions on Circuits and Systems, CAS–3:1273–1290
12. Tanaka and Saito S (1999) Neural nets and circuits. Corona Publishing (in Japanese)
13. Liu D and Michel A N (1993) Cellular neural networks for associative memories. IEEE Transactions on Circuits and Systems, CAS–40:119–121
14. Kawabata H, Zhang Z, Kanagawa A, Takahasi H and Kuwaki H (1997) Associative memories in cellular neural networks using a singular value decomposition. Electronics and Communications in Japan, III, 80:59–68
15. Szianyi T and Csapodi M (1998) Texture classification and segmentation by cellular neural networks using genetic learning. Computer Vision and Image Understanding, 71:255–270
16. Kanagawa A, Kawabata H and Takahashi H (1996) Cellular neural networks with multiple-valued output and its application. IEICE Trans. on Fundamentals of Electronics, Communications and Computer Sciences, E79-A-10:1658–1663
17. Zhang Z, Nambe M and Kawabata H (2005) Cellular neural network and its application to abnormal detection. Infromation, 8:587–604
18. Zhang Z, Akiduki T, Miyake T and Imamura T (2006) A novel design method of multi-valued CNN for associative memory. Proc. of SICE06 (in CD)
19. Namba M (2002) Studies on improving capability of cellular neural networks for associative memory and its application. PhD Thesis, Okayama Prefectural University, Japan

20. Namba M and Zhang Z (2005) The design of cellular neural networks for associative memory with multiple memory tables. Proc. 9th IEEE International Workshop on CNNA, pp.236–239

21. Kishida J, Rekeczky C,Nishio Y and Ushida A (1996) Feature extraction of postage stamps using an iterative approach of CNN. IEICE Transactions on Fundamentals of Electronics, Communications and Computer Sciences, 9:1741–1746

22. Takahashi N, Oshima K and Tanaka M (2001) Data mining for time sequence by discrete time cellular neural network. Proc. of International Symposium on Nonlinear Theory and its Applications, Miyagi, Japan, pp.271–74

23. Zhang Z, Nanba M, Kawabata H and Tomita E (2002) Cellular neural network and its application in diagnostic of automobile abnormal sound. SAE Transactions, Journal of Engines, pp.2584–2591 (SAE Paper No. 2002-01-2810)

24. Brucoli M, Cafagna D and Carnimeo L (2001) On the performance of CNNs for associative memories in robot vision systems. Proc. of IEEE International Symposium on Circuit and Systems, III:341–344

25. Tetzlaff R (ed.) (2002) Celular neural networks and their applications. World Scientific, Singapore

26. Zhang Z, Namba M, Takatori S and Kawabata H (2002) A new design method for the neighborhood on improving the CNN's efficiency. In: Tetzlaff R. (ed.) Celular neural networks and their applications, pp.609–615, World Scientific, Singapore

27. Strang G (1976) Linear algebra and its applications. Academic Press, New York

28. Japan Society for Fuzzy Theory and Systems (ed.) (1993) Fuzzy OR, Nikkan Kogyo Shimbun, Japan

29. Yanai H, Okada A, Shigemasu K, Takaki H and Yiwasaki M (ed.) (2003) Multi-variable-analysis example handbook. Asakurashoten, Japan

30. Society of Automotive Engineering of Japan (1992) Mimic sound expression for vehicle noise. Society of Automotive Engineering of Japan

31. Ulrych TJ, and Bishop TN (1975) Maximum entropy spectral analysis and autoregressive decomposition. Review of Geophysics and Space Physics, 13:180-200

5

The Wavelet Transform in Signal and Image Processing

5.1 Introduction to Wavelet Transforms

Signal analysis and image processing are very important technologies in manufacturing applications. Examples of their use include abnormal detection and surface inspection. Generally, abnormal signals, such as unsteady vibration, sound and so on have features consisting of many components, whose strength varies and whose generating time is irregular. Therefore to analyze the abnormal signals we need a time-frequency analysis method.

A number of standard methods for time-frequency analysis have been proposed and applied in various research fields [1]. The Wigner distribution (joint time-frequency analysis) and the short-time Fourier transform are typical and can be used. However, when the signal includes two or more characteristic frequencies, the Wigner distribution suffers from the confusion due to the cross terms. That is, the Wigner distribution can produce imperfect information about the distribution of energy in the time-frequency plane. The short-time Fourier transform, then, is probably the most common approach for analyzing non-stationary signals like unsteady sound and vibration. It subdivides the signal into short time segments (this is same as using a small window to divide the signal), and a discrete Fourier transform is computed for each of these. For each frequency component, however, the window length is fixed. So it is impossible to choose an optimal window for each frequency component, that is, the short-time Fourier transform is unable to obtain optimal analysis results for individual frequency components. On the other hand, the wavelet transform [2], which is a time-frequency method, does not have such problems and has some desirable properties for non-stationary signal analysis for applications in various fields, and has received much attention[3].

The wavelet transform uses the dilation b and translation a of a single wavelet function $\psi(t)$ called the *mother wavelet* (MW) to analyze all different finite energy signals. It can be divided into the continuous wavelet transform (CWT) and the discrete wavelet transform (DWT) based on the variables a

and b, which are continuous values or discrete numbers. Many famous reference books have been published[4, 5] on this topic.

However, when CWT is used in manufacturing systems as a signal analysis method, there are still two problems as follows: 1) CWT is a convolution integral in the time domain, so the amount of computation is enormous and it is impossible to analyze the signals in real time. There is still no common fast algorithm for CWT computation although it is an important technology for manufacturing systems. 2) WT can show unsteady signal features clearly in the time-frequency plane, but it cannot quantitatively detect and evaluate its features at the same time because common MW performs band pass filtering. Therefore, creating a fast algorithm and a technique for the detection and evaluation of abnormal signals is still an important subject.

Compared to CWT, a fast algorithm for DWT based on the multi-resolution analysis (MRA) algorithm has been proposed by Mallat [6]. Therefore, DWT becomes a powerful time-frequency analysis tool in the area of the data compression, de-noising and so on. DWT is a very strong tool, especially, in the area of image processing. However, DWT also has two major disadvantages, which can be shown as follows: 1) The transformed result obtained by DWT is not translation invariant [5]. This means that shifts of the input signal generate undesirable changes in the wavelet coefficients. So DWT cannot catch features of the signals exactly. 2) DWT has poor direction selection in the image [7, 8]. That is, DWT can only obtain the mixture information of $+45^o$ and -45°, although each direction information is important for the surface inspection. Therefore, how to improve the drawback DWT becomes an important subject.

We here focus on the problems shown above and show some useful improved methods as follows: 1) A fast algorithm in the frequency domain [9] for improving the CWT's computation speed. 2) The wavelet instantaneous correlation (WIC) method by using the real signal mother wavelet (RMW), which is constructed from real signals for detecting and evaluating quantitatively abnormal signals [10]. 3) Complex discrete wavelet transform (CDWT) by using the real-imaginary spline wavelet (RI-spline wavelet) for improving DWT drawbacks such as the lack of translation invariance and poor direction selection [11]. Furthermore, some applications are also given to show their effectiveness.

5.2 The Continuous Wavelet Transform

5.2.1 The Conventional Continuous Wavelet Transform

A continuous wavelet transform (CWT) maps a time function into a two-dimensional function of a and b, where a is the scale ($1/a$ denotes frequency) and b is the time translation. For a signal $f(t)$, the CWT can be written as follows:

$$w(a, b) = a^{-1/2} \int_{-\infty}^{\infty} f(t)\overline{\psi(\frac{t-b}{a})}dt, \tag{5.1}$$

where $\psi(t)$ is a mother wavelet (MW), $\overline{\psi_{a,b}(t)}$ denotes the complex conjugate of $\psi_{a,b}(t)$, and $\psi_{a,b}(t)$ stands for a wavelet basis function.

The MW $\psi(t)$ is an oscillatory function whose Fourier transform $\hat{\psi}(\omega)$ must satisfy the following admissibility condition

$$C_\psi = \int_{-\infty}^{\infty} \frac{|\hat{\psi}(\omega)|^2}{|\omega|}d\omega < \infty. \tag{5.2}$$

If this condition is satisfied, $\psi(t)$ has zero mean, and the original signal can be recovered from its transform $W(a, b)$ by the inverse transform,

$$f(t) = \frac{1}{C_\psi} \int_{-\infty}^{\infty} \int_{-\infty}^{\infty} w(a, b) a^{-1/2} \psi(\frac{t-b}{a}) \frac{dadb}{a^2}. \tag{5.3}$$

As shown in Equations 5.1 and 5.3, the CWT is a convolution integral in the time domain, so the amount of computation is enormous and it is impossible to analyze the signals in real time. In the Sect. 5.2.3, we will show a useful fast algorithm in frequency domain

Equation 5.1 shows that the wavelet transform achieves the time-frequency analysis by transforming the signal $f(t)$ into the function $w(a, b)$, which has two variables, frequency $(1/a)$ and time b. When a slow change of signal is examined, the width of the time window is enlarged by a. Conversely, when a rapid change of signal is examined, it is compressed in a. At the time that the signal change occurred, the center of the time window was removed by b.

The MW is usually classified into a real type and a complex type [12]. According to the research of the authors [13], a striped pattern is always obtained when the real MW is used since the value of $|w(a, b)|$ vibrates on the plane of time and frequency. Furthermore, the aspect of vibration of $|w(a, b)|$ changes with symmetry of the MW. This is because of the influence of the real MW's phase is considered. On the other hand, in the case of the complex MW, the value of $|w(a, b)|$ changes smoothly and a continuous pattern is obtained. Therefore, the complex MW is very useful for signal analysis. In Sect. 5.2.2, we will show an new wavelet, called the real-imaginary spline wavelet (RI-spline wavelet).

Generally, the MW has a bandpass file property. However, an abnormal signal consists of many characteristic components. So the CWT cannot detect the feature of the abnormal signal by using traditional MW. Moreover, as is shown in Equation 5.2, all functions can be used as the MW if they are functions with the characteristic that their average value is zero and the amplitude becomes zero sufficiently quickly at a distant point. Therefore, the real signal MW can be constructed by multiplying the real signal with a window function and removing the average for making it becomes zero sufficiently quickly at the distant point. In Sect. 5.2.4, we will introduce the novel real signal mother wavelet (RMW) and show an abnormal detection method by wavelet instantaneous correlation (WIC) using RMW.

5.2.2 The New Wavelet: The RI-Spline Wavelet

In this section, we first give a summary of the spline wavelets and construct a new complex wavelet, the RI-spline wavelet. Next, we examine its characteristics by using a model signal.

A. Spline Wavelet

A spline wavelet[2] with rank m can be defined as follows:

$$\psi^m(x) = \sum_{n=0}^{3m-2} q_n N_m(2x - n), \tag{5.4}$$

where the spline function $N_m(x)$ with rank m and the coefficient q_n are computed using Equations (5.5) and (5.6),

$$N_m(x) = \frac{x}{m-1}N_{m-1}(x) + \frac{m-x}{m-1}N_{m-1}(x-1), \tag{5.5}$$

$$x \in \mathbf{R},$$

$$q_n = \frac{(-1)^n}{2^{m-1}}\sum_{l=0}^{m}\binom{m}{l}N_{2m}(n+1-l), n = 0, \cdots, 3m - 2. \tag{5.6}$$

Examples of the spline wavelet are shown in Figure 5.1. Figure 5.1a shows a spline wavelet with rank $m = 5$ (spline 5) and Figure 5.1b shows a spiline wavelet with rank $m = 6$ (spline 6). Furthermore, the dual wavelet of the spline wavelet is shown in Figure 5.2. Figure 5.2a shows the dual wavelet of spline 5 and 5.2b the dual wavelet of spline 6. The spline wavelets in Figures 5.1 and 5.2 are real wavelets, having compact support in the time domain. The support of the spline wavelet $\psi^m(x)$ with rank m is $[0, 2m\text{-}1]$.

The symmetric property is an important characteristic of spline wavelets. It has an anti-symmetric property when rank m of the spline wavelet is an odd number and has a symmetric property when m is an even number. With this characteristic, the spline wavelets have a generalized linear phase, and the distortion of the reconstructed signal can be minimized.

B. RI-spline Wavelet

Here, we show a new complex wavelet, the *RI-spline wavelet*. To begin with, we use the symmetric property of the spline wavelets. We define the RI-spline wavelet as follows:

$$\psi(t) = \frac{1}{\sqrt{2}}[\psi^m(t) + i\psi^{m+1}(t)], \tag{5.7}$$

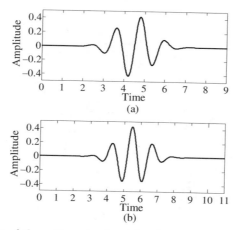

Fig. 5.1. Examples of the spline wavelet: (a) spline 5 wavelet, (b) spline 6 wavelet

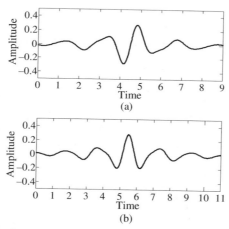

Fig. 5.2. Examples of a dual wavelet: (a) dual wavelet of the spline 5 and (b) dual wavelet of the spline 6

which has a real component when rank m is even, and an imaginary component when m is odd. In this equation $t = x - x_0$, where x_0 is the symmetrical center of $\psi^m(x)$. We define its dual wavelet as follows

$$\tilde{\psi}(t) = \frac{1}{\sqrt{2}}[\tilde{\psi}^m(t) + i\tilde{\psi}^{m+1}(t)], \tag{5.8}$$

where $\tilde{\psi}^m(t)$ and $\tilde{\psi}^{m+1}(t)$ are the dual wavelets of $\psi^m(t)$ and $\psi^{m+1}(t)$, respectively. An example of an RI-spline wavelet is shown in Figure 5.3, with the real component being the spline 6 wavelet and the imaginary component the spline 7 wavelet.

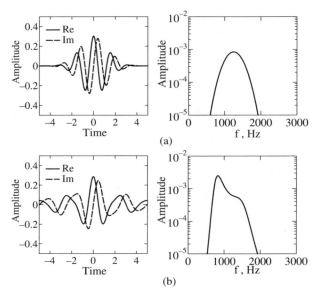

Fig. 5.3. Examples of the RI-spline wavelet: (a) the RI-spline wavelet and (b) the dual wavelet of the RI-spline wavelet

We will now analyze the properties of the RI-spline wavelet. First, we show that the RI-spline wavelet satisfies the admissibility condition given in Equation 5.2. By the symmetric property of the spline wavelet, $\psi^m(t)$ becomes an even function when m is an even number, and an odd function otherwise. That is, for any m, $\psi^m(t)\psi^{m+1}(t)$ is an odd function. Hence the result of the following integral is obvious,

$$\int_{-\infty}^{\infty} \psi^m(t)\psi^{m+1}(t)\mathrm{d}t = 0. \tag{5.9}$$

From this equation, it is clear that $\psi^m(t)$ and $\psi^{m+1}(t)$ are indeed mutually orthogonal.

The Fourier transform of the RI-spline wavelet is given as follows:

$$\hat{\psi}(\omega) = \frac{1}{2\pi} \int_{-\infty}^{\infty} \psi(t)e^{-i\omega t}\mathrm{d}t$$

$$= \frac{1}{2\pi} \int_{-\infty}^{\infty} \frac{1}{\sqrt{2}}[\psi^m(t) + i\psi^{m+1}(t)]e^{-i\omega t}\mathrm{d}t$$

$$= \frac{1}{\sqrt{2}}[\hat{\psi}^m(\omega) + i\hat{\psi}^{m+1}(\omega)]. \tag{5.10}$$

From this we obtain

$$C_\psi = \int_{-\infty}^{\infty} \frac{|\hat{\psi}(\omega)|^2}{|\omega|} d\omega$$

$$= \int_{-\infty}^{\infty} \frac{|\hat{\psi}^m(\omega)|^2}{|\omega|} d\omega + \int_{-\infty}^{\infty} \frac{|\hat{\psi}^{m+1}(\omega)|^2}{|\omega|} d\omega$$

$$= \frac{1}{2}[C_\psi^m + C_\psi^{m+1}]. \tag{5.11}$$

As the spline wavelets $\psi^m(t)$ and $\psi^{m+1}(t)$ satisfy the admissibility condition expressed in Equation 5.2, $\psi(t)$ also satisfies this condition. Therefore we may use the RI-spline wavelet to decompose and reconstruct a signal.

The RI-spline wavelets have compact support in the time domain and this can be shown easily in $[0, 2m+1]$ from the property of spline wavelets. Furthermore, it is clear from properties of spline wavelets that the RI-spline wavelets have symmetric property and a generalized linear phase.

C. Characteristics of the RI-spline Wavelet

We can define the center f^* and radius $\Delta\hat{\psi}$ of $\hat{\psi}(f)$ as one frequency window [2] as follows:

$$f^* = \frac{1}{||\hat{\psi}||_2^2} \int_{-\infty}^{\infty} f|\hat{\psi}(f)|^2 df, \tag{5.12}$$

$$\Delta\hat{\psi} = \frac{1}{||\hat{\psi}||_2}[\int_{-\infty}^{\infty} (f - f^*)^2|\hat{\psi}(f)|^2 df]^{1/2}, \tag{5.13}$$

$$||\hat{\psi}||_2^2 = \int_{-\infty}^{\infty} |\hat{\psi}(f)|^2 df.$$

In the same way, its center t^* and radius $\Delta\psi$ can be defined by making $\psi(t)$ a time window. Therefore, for time-frequency analysis, the time-frequency window by wavelet basis $\psi_{a,b}(t)$ can be written as

$$[b + at^* - a\Delta\psi, b + at^* + a\Delta\psi] \times [\frac{f^*}{a} - \frac{\Delta\hat{\psi}}{a}, \frac{f^*}{a} + \frac{\Delta\hat{\psi}}{a}]. \tag{5.14}$$

It should be noted that in this equation, the window widths $a\Delta\psi$ and $\Delta\hat{\psi}/a$ will change with scale a (or frequency) while keeping the window area $2\Delta\psi 2\Delta\hat{\psi}$ constant. Using the uncertainty principle [2], we obtain the size for the window area

$$2\pi 2\Delta\hat{\psi} 2\Delta\psi \geq 2, \quad 2\pi\Delta\hat{\psi}\Delta\psi \geq 1/2. \tag{5.15}$$

The characteristic parameters of the RI-spline wavelet and the Gabor wavelet are shown in Table 5.1. As is well-known [2], the Gabor wavelet has the best localization in time and the frequency $2\pi\Delta\hat{\psi}\Delta\psi = 0.5$. It can be

Table 5.1. Cheracterics of RI-spline and Gabor wavelets

	f^*	$\Delta\psi$ ms	$\Delta\hat{\psi}$ Hz	$2\pi\Delta\psi\Delta\hat{\psi}$
RI-spline	625	0.898	88.8	0.501
Gabor	625	0.961	82.8	0.500

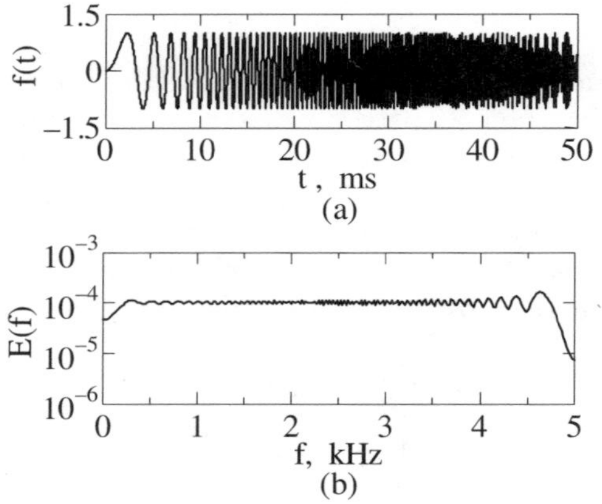

Fig. 5.4. The model signal and its energy spectrum: (a) the model signal and (b) its energy spectrum

determined from Table 5.1 that the width of the time window of the RI-spline wavelet is narrower than the Gabor wavelet and the width of the frequency window is also narrow. More importantly, from Table 5.1 the localization obtained by our RI-spline wavelet is similar to that of the Gabor wavelet. Another important property of RI-spline wavelets is that they have compact support in the time domain, which is a very desirable property.

To demonstrate the effectiveness of the RI-spline wavelet, we used a model signal $f(t)$ shown in Figure 5.4 along with its power spectrum. It has the property that each frequency component changes with time. We used 512 samples at a sampling frequency of 10 kHz, which means the Nyquist frequency $f_N = 5$ kHz.

Figure 5.5 shows a reconstructed result of the model signal using the RI-spline wavelet, where Figure 5.5a shows the reconstruction error $|f(t) - y(t)|^2$ in dB, $f(t)$ is the original signal and $y(t)$ is the reconstructed signal by using Equation 5.3, and Figure 5.5b the basis of CWT obtained by the RI-spline wavelet with six octaves and four voices. That is, the computation of the wavelet transform used four voices per octave and a frequency domain of six octave (78 Hz – 5 kHz) in which components lower than 78 Hz were cut off (78 Hz is the lowest analysis frequency in the case of 512 data samples). For such band-limited signals our RI-spline wavelet shows a better performance than that of the Gabor wavelet.

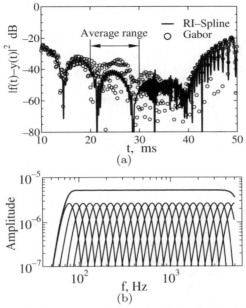

Fig. 5.5. Reconstructed error by the CWT using RI-spline and Gabor wavelets: (a) the reconstructed error and (b) the basis of the RI-spline wavelet

As is well-known, the Gabor wavelet has an infinity support, which in turn requires an infinite number of data samples. However, in real applications, we have only a finite number of data to compute Gabor wavelets. In our current experiment, the maximum support available is the number of data samples, *i.e.*, 512. That is, given finite data samples we can only approximate the Gabor wavelet, which will inevitably incur considerable errors depending on the available support. In contrast, however, because the RI-spline wavelet has a natural *compact support*, computation based on such finite data samples will result in smaller errors. Indeed, the test result in Figure 5.5a shows that the RI-spline wavelet can obtain higher precision than the Gabor wavelet. Especially, in the range of 10 ~ 20 ms, the average values of reconstruction error are -50 dB for the RI-spline wavelet and -45 dB for the Gabor wavelet, respectively. That is, the RI-spline wavelet is 5 dB better than the Gabor wavelet.

5.2.3 Fast Algorithms in the Frequency Domain

Over the last two decades, researchers have proposed some fast wavelet transform algorithms [14]. Traditionally, the *à* trous algorithm [15] and the fast algorithm in the frequency domain [16] are used. The latter is more for computation speed [17] and has the following properties: (1) It uses multiplication in the frequency domain instead of convolution in the time domain. (2) It uses one octave of the mother wavelet to obtain other mother wavelets for all octaves by down-sampling based on the self-similarity of the mother wavelet. However, this algorithm has some major problems, in particular, the computational accuracy is lower than that of the usual CWT and it is difficult to satisfy the accuracy requirement of analysis for the manufacturing systems.

We here show a fast wavelet transform (FWT), which includes the corrected basic fast algorithm and fast wavelet transform for high accuracy (FWTH) that improves the accuracy at a high computational speed. We will examine the characteristics of the FWT using a model signal and demonstrate its effectiveness.

A. Basic Algorithm for CWT

Parameters a and b in Equation 5.1 take a continuous value; however, for computational purposes, they must be digitized. Generally, when the basic scale is set to $\alpha = 2$, then $a_j = \alpha^j = 2^j$ is called octave j. For example, the Nyquist frequency f_N of the signal corresponds to the scale $a_0 = 2^0 = 1$, and the frequency $f_N/2$ corresponds to 2^1 and is referred to as one octave below f_N, or simply octave one. As for the division of the octave, we follow the method of Rioulmay [18] and divide the octave into M divisions (M voices per octave) and compute the scale as follows:

$$a_m = 2^{i/M}2^j, \tag{5.16}$$

where $i = \{0,1, \ldots,M\text{-}1\}$, $j = \{1, \ldots,N\}$, N is number of the octave, $m = i+jM$. b is digitized by setting $b = k\Delta t$, where Δt denotes the sampling interval.

As shown in Equation 5.16, the scale a_m is $2^{i/M}$ times 2^j, which expresses that the MW has a self-similarity property and the MW of the scale a_m can be calculated from the MW of the scale $2^{i/M}$ by down sampling 2^j. Therefore, we first prepare the $\psi_i(n)$ ($i=0, 1, \ldots$, M-1) for one octave from the maximum scale (the minimum analysis frequency) of analysis which corresponds to the scale $2^N 2^{-i/M}$:

$$\psi_i(n) = 2^{-\frac{N}{2}+\frac{i}{2M}}\psi(2^{-N+\frac{i}{M}}n), \tag{5.17}$$

and then calculate $\psi_m(n)$ by sampling the 2^{N-j} twice with $\psi_i(n)$,

$$\psi_m(n) = 2^{\frac{N-j}{2}}\psi_i(2^{N-j}n). \tag{5.18}$$

Finally, we rewrite Equation 5.1 as follows:

$$w(m,k) = 2^{(N-j)/2}\sum_{n=0}^{L2^j-1}\overline{\psi_i(2^{N-j}n - k)}x(n), \tag{5.19}$$

where $n = t/\Delta t$, and $L2^{j-1}$ denotes the length of the $\psi_m(n)$. Based on Equation 5.19 the number of multiplications for the CWT can be expressed as:

$$MTL \sum_{j=1}^{N} 2^{j-1} = MTL(2^N - 1), \tag{5.20}$$

where T denotes the length of the signal $x(t)$ (data length), N the number of the analysis octaves, and L the length of the $\psi_i(n)$ in $j = 1$. As shown in Equation 5.20, the amount of calculation in conventional CWT increases exponentially as the analyzing octave number N increases, because the localization of $\psi_m(n)$ becomes bad as the scale becomes large and the length $L2^{j-1}$ of $\psi_m(n)$ also increases exponentially. Moreover, the accuracy of computation becomes worse if the length of $\psi_m(n)$ is longer than the data length T, so the analysis minimum frequency (the maximum scale) will be limited by the length of the data for short data.

B. Basic Fast Algorithm for FWT

We can compute convolution in the frequency domain, for which we rewrite Equation 5.1 as follows

$$w(a,b) = a^{1/2} \int_{-\infty}^{\infty} \hat{x}(f)\overline{\hat{\psi}(af)}e^{i2\pi f b} \mathrm{d}f, \tag{5.21}$$

where $\hat{x}(f)$, $\hat{\psi}(af)$ are Fourier transforms of $x(t)$ and $\psi(t/a)$, respectively, and $\overline{\hat{\psi}(af)}$ denotes the complex conjugate of $\hat{\psi}(af)$. In addition a basic fast algorithm (BFA) of wavelet transform in the frequency domain has been developed[6].

As was done above, we first compute one octave of $\hat{\psi}_i(n)$ from f_N, the minimum scale (analysis maximum frequency), $a=2^{i/M}$ $(j=0)$,

$$\hat{\psi}_i(n) = 2^{\frac{i}{2M}} \hat{\psi}(2^{\frac{i}{M}} n), \tag{5.22}$$

where $n=f/\Delta f$, and $\Delta f=1/T$; Δf denotes the frequency interval. We then use the self-similarity of MW to obtain another MW for all octaves as follows

$$\hat{\psi}_m(n) = 2^{\frac{j}{2}} \hat{\psi}_i(2^j n). \tag{5.23}$$

Consequently, the $\hat{x}(f)\overline{\hat{\psi}(af)}$ in (5.21) can be rewritten as follows

$$w(m,n) = 2^{\frac{j}{2}} \hat{x}(n)\overline{\hat{\psi}_i(2^j n)}. \tag{5.24}$$

thus $w(m,k)$, which is a discrete expression of $w(a,b)$, can be obtained by using the inverse Fourier Transform about k as follows

$$w(m,k) = 2^{\frac{j}{2}} \sum_{n=0}^{T} w(m,n)e^{i2\pi \frac{nk}{T}}. \tag{5.25}$$

We now consider the number of multiplications for the BFA based on Equation (5.25). Roughly, $L\log_2 L$ multiplications are required for one reverse

Fourier transform and T multiplications for $\hat{x}(n)$, and $\overline{\hat{\psi}_m(n)}$. That is, in order to calculate $w(m, k)$, we need the number of multiplications:

$$MN(T + T\log_2 T) = MNT(1 + \log_2 T). \tag{5.26}$$

As shown above, the amount of calculations of the FWT based on the BFA is different from the CWT, and is sensitive to the data length T. Moreover, the localization of the MW in the frequency domain becomes better as the scale becomes larger (analysis frequency becomes small). So the analysis range available in the FWT will be larger than that in the conventional WT. Theoretically, the frequency range of the FWT can be analyzed until the length of $\psi_m(n)$ becomes one piece. For example, the FWT is analyzable to ten octaves (4.9-5.0 kHz) with $T = 512$ and symmetrical boundary condition. However, the CWT is analyzable only to six octaves (78 Hz-5.0 kHz) under the same conditions.

However, the FWT has a higher reconstructed error (RE) than that obtained by CWT. Next, we will show techniques to improve accuracy.

C. Improving Accuracy

In order to compare the computational accuracy between CWT and FWT, we used the model signal $f(t)$ shown in Figure 5.4 along with its power spectrum. It has the property that each frequency component changes with time, and has 512 samples at a sampling frequency of 10 kHz. We use the RI-spline wavelet shown in Sect. 5.2.2 as the MW, and first perform a wavelet transform of the original signal $f(t)$ in order to get $W(a, b)$, we then obtain the reconstructed signal $y(t)$ from the inverse wavelet transform. Figure 5.6 shows the reconstructed error $|f(t) - y(t)|^2$ in dB. Figure 5.6a shows the result obtained from the CWT with six analysis octaves (78 Hz-5 kHz) and Figure 5.6b shows the result obtained from the FWT based on the BFA with ten analysis octaves (4.9 Hz-5 kHz). Both computations used four voices per octave.

It is clear by comparing Figures 5.6a and 5.6b that the CWT has a better performance than the FWT. In the case of the CWT about -40 dB of RE is obtained when removing the low frequency domain and the high frequency domain, but in the case of the FWT only about -20 dB of RE is obtained over the entire frequency domain. This is because all MWs in the case of the FWT based on the BFA are obtained from the MWs near the Nyquist frequency, which have less data and whose calculation accuracy is low, although they have good localization in the time domain.

In order to improve the computational accuracy of the FWT, we use MWs whose frequencies are two octaves lower than the Nyquist frequency. This results in the corrected basic fast algorithm (CBFA). In this case, the length of the MWs in a time domain is extended four times from the length of the MWs near the Nyquist frequency. If one fourth of the beginning of the data is used after carrying out the Fourier transform of the MWs with four times the data length, the MWs obtained have the same length (localization) as the

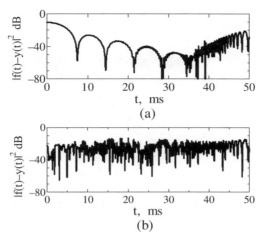

Fig. 5.6. RE by using the CWT and FWT based on the BFA: (a) the reconstruction error by CWT, and (b) the reconstruction error by FWT

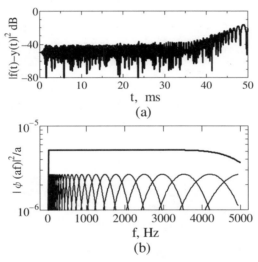

Fig. 5.7. RE improved by using the FWT based on the CBFA and wavelet bases with ten octave, four divided: (a) the improved reconstruction error and (b) the basis of the FWT

MWs near the Nyquist frequency. Figure 5.7 shows the result obtained from the FWT using the CFBA. Figure 5.7a shows the RE and Figure 5.7b shows the basis system constructed by $\hat{\psi}_m(n)$. As shown in Figure 5.7, the FWT based on CBFA shows a good performance. The RE obtained is lower than -40 dB over a wide frequency range and the accuracy is better than the result

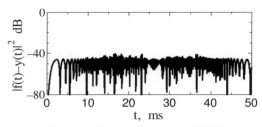

Fig. 5.8. RE by using the FWTH

of CWT shown in Figure 5.6a. This method does not have any influence on the computational speed because the parameters in Equation 5.26 have not changed. However, in the high frequency domain, the problem that the RE is larger still remains. This is because when the frequency approaches f_N, the amplitude value of $\sum |\hat{\psi}(af)|/a$ (shown in Figure 5.7(b)) becomes small.

In order to obtain a high degree of accuracy in the high frequency domain, we use up-sampling by using L-spline interpolation. We assume that the sampling frequency does not change after the data is interpolated although the number of samples increases twofold, so that the frequency of each frequency component falls by half. Therefore, the influence due to the reduction of the amplitude value of $\sum |\hat{\psi}(af)|/a$ near f_N is avoided. This method is called fast wavelet transform for high accuracy (FWTH). Figure 5.8 shows the result obtained by using FWTH. As shown in the figure, the RE obtained is lower than -40 dB over the entire frequency range. However, higher accuracy is at the expense of computational speed because the data length has been doubled.

D. Computational Speed

The relative calculation time (RCT) for our methods is shown in Figure 5.9. The analysis data length is 512, the number of analysis octaves is six and each octave has been divided into four voices. The RCT of each method is expressed relative to traditional CWT where the calculation time is set at 100. As shown in Figure 5.9, the computation time of FWT using CBFA is only 3.33. For FWTH, the computation time is only 7.62 although the data length is double in order to compute FWT with high accuracy. Based on the discussion above, we may conclude that the proposed FWT is indeed effective for improving the computation accuracy at a high computation speed. That the RCT of FWT and traditional CWT changes with the increase in the number of analysis octaves was shown in Figure 5.10. That the ratio with computation quantity (RCQ) based on Equations 5.20 and 5.26 changes with the increase in the number of analysis octaves was also shown in Figure 5.10 for comparison. The value of both RCT and RCQ are expressed with the ratio setting the value of CWT in five octaves as 100. As shown in Figure 5.10, the change of RCT about traditional CWT and FWT is well in agreement with

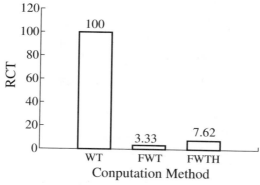

Fig. 5.9. Ratio of the computation time (RCT)

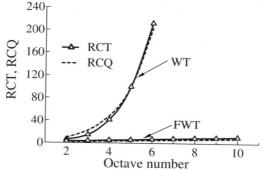

Fig. 5.10. RCT and RCQ changes with octave number

RCQ. The calculation time of FWT increase is small, and oppositely, the calculation time of CWT increases abruptly by the increase in the number of analysis octaves. This is well in agreement with the discussion above, and it is demonstrated that FWT can be adapted for a wider analysis frequency range than traditional CWT. Moreover, the change of RCT is approximately the same as RCQ with a change of data length T and a voice number M, so RCT can be predicted using RCQ.

5.2.4 Creating a Novel Real Signal Mother Wavelet

As is shown in Sect. 5.2.1, the MW must satisfy the admissibility condition shown in Equation 5.2. Actually, this condition can be simplified to the next equation when $\psi(t)$ tends to zero and t approaches infinity.

$$\int_{-\infty}^{\infty} \psi(t)\mathrm{d}t = 0. \tag{5.27}$$

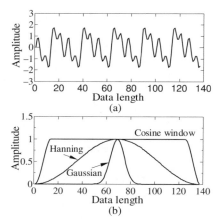

Fig. 5.11. Model signal and window functions: (a) the model signal and (b) the window functions

Moreover, all functions can be used as the MW if they are functions with the characteristic that their average value is zero and the amplitude becomes zero sufficiently quickly at a distant point. Therefore, a real signal mother wavelet (RMW) can be constructed by multiplying the real signal with a window function and removing the average for making it become zero sufficiently quickly at the distant point. Here, how selection of the window function and construction of the complex RMW is performed should be noted.

A. Selecting the Window Function for the RMW

We first examine the influence of the window function in the construction of the real RMW by taking the case of a model signal consisting of three sine waves with 400 Hz, 800 Hz and 1600 Hz.

$$f(t) = \sin(800\pi t) + 0.7\sin(1200\pi t) + 0.7\sin(3200\pi t), \qquad (5.28)$$

where t denotes time. Figure 5.11a shows the model signal generated by Equation 5.28. The window functions, cosine window, Hanning window and Gaussian function that are usually well used are also shown in Figure 5.11b.

The cosine window is the window function that is the multiplication of the cosine wave by $1/10$ the portion of the signal length T to the both ends of the signal. The Hanning window and the Gaussian function are given by Equations 5.29 and 5.30, respectively.

$$W_H(k) = \begin{cases} \frac{1}{2}(1 + \cos(\frac{\pi k}{\tau_m})) & |k| < \tau_m \\ 0 & |k| > \tau_m, \end{cases} \quad \tau_m = T/2, \qquad (5.29)$$

$$W_G(k) = \frac{1}{\sqrt{2\pi}}e^{(k-\mu)^2/2}, \quad \mu = T/2 \qquad (5.30)$$

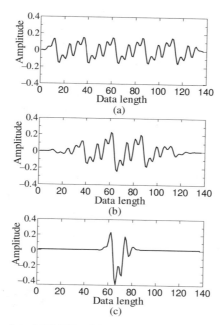

Fig. 5.12. Example of real RMWs: (a) made by a cosine window, (b) made by a Hanning window, and (c) made by a Gaussian window.

The window's width of the three kinds of window functions shown above serves as the cosine window, the Hannning window, and the Gaussian function in the order of the width. On the other hand, in the order of the smoothness of the windows, it is the order of the Gaussian function, the Hannning window, and the cosine window.

Figure 5.12 shows examples of real RMW $\psi^R(t)$ which are constructed by carrying out the multiplication of the window function to the model signal shown in Figure 5.11, and subtracting this average value. Furthermore, those power spectrums are shown in Figure 5.13, where according to Figures 5.12a and 5.13a the results are obtained by cosine window. In Figures 5.12b and 5.13b they are obtained by the Hanning window, and Figures 5.12c and 5.13c by the Gaussian function. Moreover, the norm of each RMW $||\psi^R||$ is set to 1,

$$||\psi^R|| = \left[\int_{-\infty}^{\infty} \psi^R(t)^2 dt\right]^{1/2} = 1, \tag{5.31}$$

From Figures 5.12 and 5.13 it is clear that the RMW obtained by the cosine window has large window width in the time domain and high frequency resolution. However, the vibration power spectrum was obtained because the smoothness of the window function was inadequate. On the other hand, the

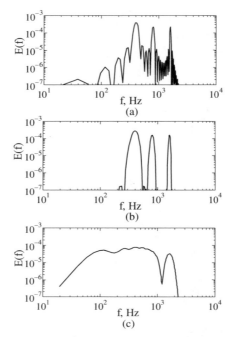

Fig. 5.13. Frequency spectrum $|\hat{\psi}^R(f)|$ of real RMW: (a) obtained by using a cosine window, (b) by a Hanning window, and (c) by a Gaussian window

RMW by the Gaussian function has higher time resolution since it includes only the local information in the time domain although the frequency resolution is lower so it cannot recognize the peaks of 400Hz and 800Hz. As a comparison, the RMW by the Hanning window has good time and frequency resolution in three kinds of window functions; it is the optimal window function in this research. Therefore, the Hannning window has been adopted therein.

B. Constructing the Complex RMW

Usually, a complex RMW can be expressed by the following formula

$$\psi(t) = \psi^R(t) + j\psi^I(t), \tag{5.32}$$

where, j is the unit of the imaginary and $\psi^R(t)$ and $\psi^I(t)$ are the real component and imaginary component of the $\psi(t)$, respectively. Generally, the Fourier transform $|\hat{\psi}^R(f)|$ of the real wavelet function $\psi^R(t)$ has a symmetrical frequency spectrum in the positive and negative frequency domains, as shown in Figure 5.13. On the other hand, the Fourier transform $|\hat{\psi}(f)|$ of the complex wavelet function $\psi(t)$ exists only in a positive frequency domain, and is $|\hat{\psi}(f)| = 0$ in the negative domain. For fulfilling this characteristic, the required and sufficient condition is that $\psi^I(t)$ is a Hilbert pair of $\psi^R(t)$[19].

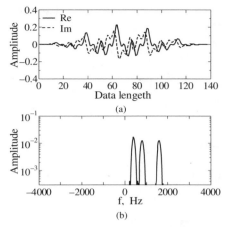

Fig. 5.14. Example of complex RMW $\psi(t)$ and its frequency spectrum $|\hat{\psi}(f)|$: (a) complex RMW and (b) frequency spectrum

Then, it tries to construct a complex RMW using the frequency characteristic of the complex wavelet function. The procedure can be summarized as follows: (1) Carrying out a Fourier transform of the real RMW $\psi^R(t)$ and obtaining its frequency spectrum $\hat{\psi}^R(f)$. (2) In the negative frequency domain, $\hat{\psi}^R(f)$ is set to 0, in the positive frequency domain, $\hat{\psi}^R(f)$ is set to $\sqrt{2}\hat{\psi}^R(f)$ and carrying out reverse Fourier transform. In procedure (2), in order to calculate the Hilbert transform of $\psi^R(t)$ the multiplication of $\hat{\psi}^R(f)$ and 2 in the positive frequency domain is usually used [19]. However, in order to have the same power spectrum ($||\psi||$=1) between the real and the complex RMWs, the multiplication of $\hat{\psi}^R(f)$ and $\sqrt{2}$ is used.

Figure 5.14a shows an example of the complex RMW $\psi(t)$ constructed by using the real RMW $\psi^R(t)$ that was shown in Figure 5.12. Figure 5.14b shows its frequency spectrum $|\hat{\psi}(f)|$. The power spectrum $E(f)$ obtained from the frequency spectrum shown in Figure 5.14b is the same as Figure 5.13b, and that the norm of $\psi(t)$ becomes $||\psi||$=1 is confirmed.

C. Definition of WIC by using RMW

Figure 5.15 shows a calculation result of the CWT, in which the real RMW shown in Figure 5.12b is considered as the signal. For the MW, in Figure 5.15a the RI-spline wavelet [13] that is well used for the default CWT is used. As a comparison, in Figure 5.15b the complex RMW shown in Figure 5.14a is used and in Figure 5.15c the real RMW shown in Figure 5.12b is used. Moreover, the horizontal axis of Figure 5.15 shows the time. The vertical axis of Figure 5.15a shows the frequency, the vertical axis of Figures 5.15b and 5.15c show the scale a and the shade level shows the amplitude of $|w(a,b)|$. A calculation sampling

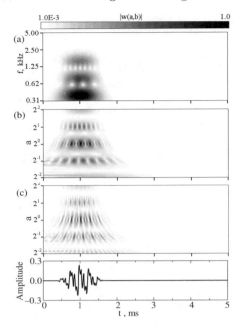

Fig. 5.15. Wavelet transform by using the RI-spline wavelet and RMWs

interval (0.1 ms) is used for the horizontal axis and a division of each octave into 32 voices with the log scale is used for the vertical axis.

From Figure 5.15a it is clear that three patterns consisting mainly of 400 Hz, 800 Hz, and 1600 Hz were obtained, and the pattern centering on 400Hz has appeared comparatively strongly when the RI-spline wavelet is set to the MW. This is well in agreement with the characteristic of the original signal. Compared with Figure 5.15a, in Figures 5.15b and 5.15c, the pattern consisting mainly of the scale $a = 1$ and 1 ms has appeared strongly. This is because the RMW has completely the same components as the analysis signal in the scale $a = 1$ and 1ms, that is, the RMW has strong correlation with the analysis signal in the scale $a = 1$ and 1ms.

In the scale $a = 1$, amplitude of $|w(a,b)|$ changes strangely when RMW is moved to just over or just below 1 ms. Moreover, in Figures 5.15b and 5.15c, comparatively weak patterns exist around scale $a = 2, 0.5$. This is because the components of the RMW, for example, the component of 800 Hz becomes 1600 Hz if twice 800 Hz, or becomes 400 Hz if it is 0.5 times, and the components of 1600 Hz and 400 Hz have a correlation with the same components of the RMW. In addition, the differences between Figures 5.15b and 5.15c are the striped pattern obtained by the real RMW and continuation pattern obtained by the complex RMW.

Then, the value $|w(a, b)|$ obtained by the RMW in the scale $a = 1$ is defined by the wavelet instantaneous correlation value $R(b)$ and is shown as follows:

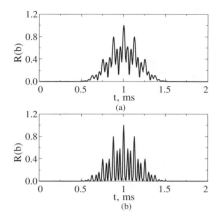

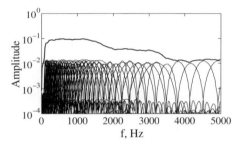

Fig. 5.16. $R(b)$ obtained by complex RMW and real RMW, respectively: (a) $R(b)$ obtained by complex RMW, (b) obtained by real RMW

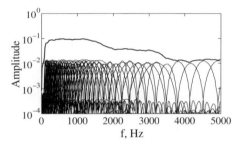

Fig. 5.17. Basis made by the RMW filter bank shown in Figure 5.14b

$$R(b) = |w(a = 1, b)|. \qquad (5.33)$$

Furthermore, Figure 5.16 shows the $R(b)$ obtained from Figures 5.15b and 5.15c at the scale $a = 1$ and plotting them in time t ($t = b$). Figure 5.16a is obtained by the complex RMW, and Figure 5.16b by the real RMW. As shown in Figure 5.16, $R(b) = 1.0$ can be obtained in 1 ms since the RMW is completely the same as the components of the signal, that is, the generation time and the strength of the signal can be extracted simultaneously by the amplitude of $R(b)$. Furthermore, in the case of the real RMW shown in Figure 5.16b, $R(b)$ has an oscillating phenomenon. On the other hand, in the case of the complex RMW shown in Figure 5.16a, the oscillating phenomenon of $R(b)$ can be improved. Therefore, the complex RMW is very useful for this study and it will be used in the following examples.

Figure 5.17 shows that the filter bands of $\hat{\psi}_{a,b}(f)$ defined from the RMW $\hat{\psi}(f)$ are arranged in order in a frequency domain, where each octave is divided into four voices. As shown in Figure 5.17, the characteristic of the base constructed by the filter bands of $\hat{\psi}_{a,b}(f)$ is not good, although the base is redundant and perfect. This is because $\hat{\psi}(f)$ contains two or more characteristic components (as shown in Figure 5.14b). That is, the reconstruction accuracy cannot be guaranteed although the reverse transform using the RMW exists. The purpose is not to reconstruct the signal but to extract the components that are similar to the RMW and are embedded in the analysis signal. As shown in Figure 5.16, the generating time and the strength of the components that are same as the RMW in the analysis signal can be extracted simultaneously if the $R(b)$ proposed in this study is used. We believe that our study goal can be attained by using the RMW.

5.3 Translation Invariance Complex Discrete Wavelet Transforms

5.3.1 Traditional Discrete Wavelet Transforms

In Equation 5.1 shown in Sect. 5.2.1, when we use variable a and b such that $a = 2^{-j}$, $b = 2^{-j}k$ with two positive numbers j and k, the wavelet transform is called discrete wavelet transform (DWT), and letting $w(j, k)$ be equal to d_k^j, the DWT can be shown as follows:

$$d_{j,k} = \int_{-\infty}^{\infty} f(t)\overline{\psi_{j,k}(t)}\mathrm{d}t, \qquad (5.34)$$

$$\psi_{j,k}(t) = 2^{j/2}\psi(2^j t - k),$$

where $\psi_{j,k}(t)$ denotes the wavelet basis functions obtained from an original wavelet $\psi(t)$, and $\overline{\psi_{j,k}(t)}$ expresses the complex conjugate of $\psi_{j,k}(t)$. k denotes time and j is called the level (or 2^j is called the scale). In the case of the DWT, $\psi(t)$ must satisfy the following bi-orthogonal condition for signal reconstructability:

$$\int_{-\infty}^{\infty} \overline{\psi_{j,k}(t)}\tilde{\psi}_{l,n}(t)\mathrm{d}t = \delta_{j,l}\delta_{k,n}, \qquad (5.35)$$

where $\tilde{\psi}(t)$ is called a dual wavelet of $\psi(t)$, $\tilde{\psi}_{l,n}(t)$ is the dual wavelet basis function derived from $\tilde{\psi}(t)$. Generally, in the case of the orthogonal wavelet, $\psi(t) = \tilde{\psi}(t)$. However, in the case of the non-orthogonal wavelet, $\psi(t) \neq \tilde{\psi}(t)$, which is clear from Equation 5.35. Therefore, we need to find a dual wavelet of $\psi(t)$, such as the spline wavelet and its dual wavelet, which has been found by Chui and Wang [2]. The original signal can then be reconstructed by

$$f(t) = \sum_j \sum_k d_{j,k}\tilde{\psi}_{j,k}(t). \qquad (5.36)$$

Different from CWT, a very efficient fast algorithm for achieving DWT by using the multi-resolution analysis (MRA) has been proposed by Mallat[6]

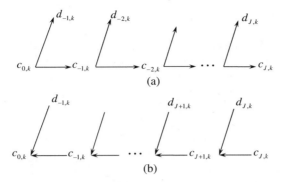

(a)

(b)

Fig. 5.18. Mallat's fast algorithm for DWT: (a) decomposition tree, and (b) reconstruction tree

(see Figure 5.18). As shown in Figure 5.18, generally, Mallat's fast algorithm first starts from level 0, where the signal $f(t)$ is approximated by $f_0(t)$, and the signal is decomposed by the following formula:

$$f_0(t) = \sum_k c_{,k} \phi(t - k), \quad k \in \mathbf{Z}, \tag{5.37}$$

where $\phi(t)$ means a scaling function, $c_{,k}$ digital data of the signal $f_0(t)$.

Then following the decomposition tree shown in Figure 5.18a, the decomposition can be calculated by the following equations:

$$c_{j,k} = \sum_l a_{l-2k} c_{j-1,l}, \tag{5.38}$$

$$d_{j,k} = \sum_l b_{l-2k} c_{j-1,l}, \tag{5.39}$$

where the sequences $\{a_k\}$ corresponding to scaling function $\phi(t)$, and $\{b_k\}$ corresponding to wavelet $\psi(t)$ denote the decomposition sequences. Furthermore, following the reconstruction tree shown in Figure 5.18b, the inverse transformation can be calculated by the following equations:

$$c_{j,k} = \sum_l (p_{k-2l} c_{j-1,l} + q_{k-2l} d_{j-1,l}), \tag{5.40}$$

where the sequences $\{p_k\}$ and $\{q_k\}$ denote the reconstruction sequences . The decomposition and reconstruction sequences are explained in several references, e.g. the sequences of Daubechies wavelets are shown in [4] and spline wavelets are shown in [2], respectively.

However, the DWT computed by the MRA algorithm has a translation variance problem[5]. This problem hinders the DWT from being used in wider fields. Currently, successful applications of DWT are restricted to image compression, *etc.*

Some methods have been proposed to create translation invariant DWT. Kingsbury [7, 8] proposed a complex wavelet transform, the dual-tree wavelet transform (DTWT), which achieves approximate translation invariance and takes only twice as much computational time as the DWT for one dimension (2^m times for m dimensions). However, the major drawback of Kingsbury's approach is that, in the process of creating a half-sample-delay, level -1 decomposition results cannot be used for complex analysis. So it is difficult to use Kingsbury's DTWT for signal and image processing. Fernandes et al. [20] proposed a new framework for the implementation of complex wavelet transforms (CWTs), which uses a mapping filter for obtaining Hilbert transform pairs of input data and twice traditional DWT for obtaining real and imaginary wavelet coefficients, respectively. However, in the case of one dimension (1-D), the computational time of the CWTs is longer than that of the DTWT due to using twice that of the mapping filter (mapping and inverse-mapping) and twice that of the DWT. The same is true in the case of two dimension (2-D).

On the other hand, for the complex discrete wavelet transform (CDWT), Zhang et al. [11] proposed a new complex wavelet: the real-imaginary spline wavelet (RI-spline wavelet) and a coherent dual-tree algorithm instead of the framework of Kingsbury's DTWT. Furthermore, this method has been applied to de-noising and image processing and so on, and its effectiveness has been shown. Therefore, we will introduce the CDWT in next sections.

5.3.2 RI-spline Wavelet for Complex Discrete Wavelet Transforms

A. RI-spline Wavelet for DWT

In the Sect. 5.2.2, a complex wavelet, the *RI-spline wavelet* was defined as follows:

$$\psi(t) = \frac{1}{\sqrt{2}}[\psi^m(t) + i\psi^{m+1}(t)], \qquad (5.41)$$

which has a real component when rank m is even (m_e), and an imaginary component when $m + 1$ is odd (m_o). Dual wavelet has also been defined as follows:

$$\tilde{\psi}(t) = \frac{1}{\sqrt{2}}[\tilde{\psi}^m(t) + i\tilde{\psi}^{m+1}(t)], \qquad (5.42)$$

where $\tilde{\psi}^m(t)$ and $\tilde{\psi}^{m+1}(t)$ are the dual wavelets of $\psi^m(t)$ and $\psi^{m+1}(t)$, respectively.

We here simply use the following notation:

$\psi^R(t)$: the real component of the RI-spline wavelet.
$\psi^I(t)$: the imaginary component of the RI-spline wavelet.
$\tilde{\psi}^R(t)$: the real component of the dual RI-spline wavelet.
$\tilde{\psi}^I(t)$: the imaginary component of the dual RI-spline wavelet.
$N^R(t)$: the real component of the RI-spline scaling function.

$N^I((t)$: the imaginary component of the RI-spline scaling function.
$\psi^{m_e}(t)$: the m_e (the rank is an even number) spline wavelets.
$\psi^{m_o}(t)$: the m_o (the rank is an odd number) spline wavelets.
$N_{m_e}(t)$: the m_e spline scaling function.
$N_{m_o}(t)$: the m_o spline scaling function.

Using these notations we show the RI-spline wavelet and its scaling functions as follows:

$$\psi(t) = \psi^R(t) + j\psi^I(t),$$
$$\psi_R(t) = (-1)^{(m_e-2)/2}||\psi^{m_e}||^{-1}\psi^{m_e}(t + m_e - 1), \tag{5.43}$$
$$\psi_I(t) = (-1)^{(m_o+1)/2}||\psi^{m_o}||^{-1}\psi^{m_o}(t + m_o - 1),$$

$$N^R(t) = N_{m_e}(t - m_e/2),$$
$$N^I(t) = N_{m_o}(t - (m_o - 1)/2), \tag{5.44}$$

where Equations 5.43 and 5.44 imply phase adjustment. The normalization of the wavelets is conducted as follows:

$$\langle \psi^R, \psi^I \rangle = 0,$$
$$||\psi^R|| = ||\psi^I|| = 1, \tag{5.45}$$

where $||\psi^R||$ or $||\psi^I||$ is definite as is Equation 5.31 and $\langle \psi^R, \psi^I \rangle$ is definite as follows:

$$\langle \psi^R, \psi^I \rangle = \int_{-\infty}^{\infty} \psi^R(t)\bar{\psi}^I(t)\mathrm{d}t = 0. \tag{5.46}$$

In the case of DWT, the $\psi(t)$ must satisfy the bi-orthogonal condition shown in Equation 5.35 for signal reconstructability. Fortunately, that the RI-spline wavelet satisfies the bi-orthogonal condition was shown in [13]. In other words, the RI-spline wavelet can be used as a mother wavelet of the discrete wavelet transform.

Figure 5.19a shows the basis of the DWT using the RI-spline wavelet and Figure 5.19b shows one example of the frequency window (filter bank) of $\hat{\psi}(f)$, $\hat{\tilde{\psi}}(f)$ and $\hat{\psi}(f)\hat{\tilde{\psi}}(f)$. As is shown in Figure 5.19a, a complete basis is constructed very well by using $\hat{\psi}(f)\hat{\tilde{\psi}}(f)$, and it is different from the CWT shown in Figure 5.7b. In the case of the DWT, a signal is analyzed with the octaves are shown in Equation 5.34, and it is therefore necessary to make the dual wavelet $\hat{\tilde{\psi}}(f)$ a pair of $\hat{\psi}(f)$ because the width of the frequency window of $\hat{\psi}(f)$ is narrower as shown in Figure 5.19.

The reconstructed error of the model signal is shown in Figure 5.20, where the RI-spline wavelet is used as the MW and Equations 5.34 and 5.36 have been used to calculate the DWT. Figure 5.20 shows that the reconstructed error is less than -45 dB over all frequencies. This is better than the CWT especially in the high frequency domain. We think it is because the amplitude of the basis system shown in Figure 5.19a is also fixed near f_N, and is different from the CWT shown in Figure 5.7b.

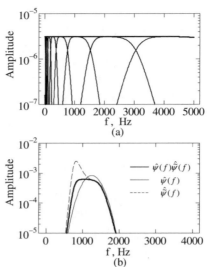

Fig. 5.19. Basis of the discrete wavelet transform and filter band: (a) basis of discrete wavelet transform, and (b) the wavelets as a filterbank

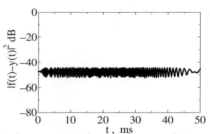

Fig. 5.20. Reconstructed error using the discrete wavelet transform with ten octaves

B. RI-spline Wavelet for CDWT

We denote the decomposition sequences of $\psi^R(t)$ as $\{a_k^R\}$ and $\{b_k^R\}$, and those of $\psi^I(t)$ as $\{a_k^I\}$ and $\{b_k^I\}$. We also denote the decomposition sequences of $\psi^{m_e}(t)$ as $\{a_k^{m_e}\}$ and $\{b_k^{m_e}\}$, and those of $\psi_{m_o}(t)$ as $\{a_k^{m_o}\}$ and $\{b_k^{m_o}\}$. Using this notation the decomposition sequences of the RI-spline wavelet are expressed as follows:

$$a_k^R = \sqrt{2}a_{k+m_e/2}^{m_e},$$
$$b_k^R = (-1)^{m_e/2+1}||\psi^{m_e}||\sqrt{2}b_{k+3m_e/2-2}^{m_e}, \tag{5.47}$$

$$a_k^I = \sqrt{2}a_{k+(m_o-1)/2}^{m_o},$$
$$b_k^I = (-1)^{(m_o+1)/2}||\psi^{m_o}||\sqrt{2}b_{k+3(m_o-1)/2}^{m_o}. \tag{5.48}$$

We denote the reconstruction sequences of $\psi^R(t)$ as $\{p_k^R\}$ and $\{q_k^R\}$, and those of $\psi^I(t)$ as $\{p_k^I\}$ and $\{q_k^I\}$. We also denote the reconstruction sequences of the

m_e spline wavelet as $\{p_k^{m_e}\}$ and $\{q_k^{m_e}\}$, and those of the $m = m_o$ spline wavelet as $\{p_k^{m_o}\}$ and $\{q_k^{m_o}\}$. Using this notation the reconstruction sequences of the RI-spline wavelet are expressed as follows:

$$
\begin{aligned}
p_k^R &= (\sqrt{2})^{-1} p_{k+m_e/2}^{m_e}, \\
q_k^R &= (-1)^{m_e/2+1} (\|\psi^{m_e}\| \sqrt{2})^{-1} q_{k+3m_e/2-2}^{m_e},
\end{aligned}
\tag{5.49}
$$

$$
\begin{aligned}
p_k^I &= (\sqrt{2})^{-1} p_{k+(m_o-1)/2}^{m_o}, \\
q_k^I &= (-1)^{(m_o+1)/2} (\|\psi^{m_o}\| \sqrt{2})^{-1} q_{k+3(m_o-1)/2}^{m_o}.
\end{aligned}
\tag{5.50}
$$

In Equations 5.47, 5.48, 5.49 and 5.50, we omit the normalization of wavelets in each level.

5.3.3 Coherent Dual-tree Algorithm

A. Creating a Half-sample Delay Using Interpolation

As shown in Sect. 5.3.1, generally, Mallat's fast algorithm for DWT first starts from level 0, where the signal $f(t)$ is approximated by $f_0(t)$, and the signal is decomposed by the following formula:

$$
f_0(t) = \sum_k c_{0,k} \phi(t - k), \quad k \in \mathbf{Z}.
\tag{5.51}
$$

In this equation, $\phi(t)$ means a scaling function, and c_k^0 the digital data of the signal $f_0(t)$.

Usually, in the spline wavelet, as the scaling function $N_m(t)$ is not orthogonal, and the signal $f(t)$ is approximated by $f_0(t)$ using the following interpolation:

$$
f_0(t) = \sum_k f(k) L_m(t - k), \quad k \in \mathbf{Z}.
\tag{5.52}
$$

The fundamental spline $L_m(t)$ of rank m is defined as

$$
L_m(t) = \sum_k \beta_k^m N_m(t + \frac{m}{2} - k), \quad k \in \mathbf{Z},
\tag{5.53}
$$

which has the interpolation property $L_m(k) = \delta_{k,0}$, $k \in \mathbf{Z}$. β_k^m is the coefficient of Equation (5.53) and $\delta_{k,0}$ is defined as follows:

$$
\delta_{k,j} = \begin{cases} 0, & k = j \\ 1, & k \neq j \end{cases}.
\tag{5.54}
$$

Using Equations 5.52 and 5.53, we obtain the following equations

$$
f_0(t) = \sum_k f(k) L_m(t - k), \quad k \in \mathbf{Z}
\tag{5.55}
$$

$$
= \begin{cases} \sum_k c_{0,k} N_m(t-k) & m = m_e,\ k \in \mathbf{Z} \\ \sum_k c_{0,k} N_m\left(t + \dfrac{1}{2} - k\right) & m = m_o,\ k \in \mathbf{Z} \end{cases} ,
$$

$$
c_{0,k} = \begin{cases} \sum_l f(l) \beta^m_{k+m/2-l} & m = m_e,\ l \in \mathbf{Z} \\ \sum_l f(l) \beta^m_{k+(m-1)/2-l} & m = m_o,\ l \in \mathbf{Z} \end{cases} . \tag{5.56}
$$

As shown in Equations 5.55 and 5.56, when m is m_e, $f_0(x)$ becomes the standard form expressed in Equation 5.51. However, when m is m_o, a half-sample-delay from the case where m is m_e occurs in $c_{0,k}$.

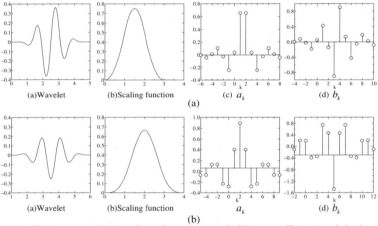

Fig. 5.21. Spline wavelet, scaling function and filter coefficients: (a) the case of $m = 3$ spline wavelet, (b) $m = 4$ spline wavelet

Figure 5.21 shows an example of spline wavelets and their filters. As shown in the figure, if one uses the m_e spline wavelet as a real component and the m_o spline wavelet as an imaginary component, then there is a half-sample-delay between the two filters. Therefore in the CDWT calculation, one must provide a half-sample-delay for two filters in level 0. Fortunately, as shown above, this half-sample-delay can be easily achieved in the process of interpolation calculation when the m_e and m_o spline scaling functions are used. However, the coefficient β^m_k in Equation 5.53 is very difficult to calculate in the case where m is m_o [2]. In order to calculate this coefficient, we show a new synthetic-interpolation function, which is defined as follows:

$$
N_s(t) = \sum_k K^R_k N_R(t-k) + \sum_k K^I_k N_I(t-k),\ k \in \mathbf{Z}. \tag{5.57}
$$

In Equation (5.57), it is necessary for $N_s(t)$ to be symmetric around the origin. It is also necessary for the energy of the input signal to be evenly shared in the real component $\sum K_k^R N_R(t-k)$ and the imaginary component $\sum K_k^I N_I(t-k)$, except near the Nyquist frequency. The sequences K_k^R and K_k^I designed so that they satisfy these conditions are shown following Equations:

$$K_k^R = \begin{cases} 1 & k = 0 \\ 0 & \text{otherwise} \end{cases}, \tag{5.58}$$

$$K_k^I = \begin{cases} \frac{l_{-k}}{T} & -5 \leq k \leq -1 \\ \frac{l_0}{T} & 0 \leq k \leq 1 \\ \frac{l_{k-1}}{T} & 2 \leq k \leq 6 \\ 0 & \text{otherwise} \end{cases}, \tag{5.59}$$

$$l_k = \begin{cases} 4.5 & k = 0 \\ (-0.55)^k & 1 \leq k \leq 5 \end{cases},$$

$$T = 2 \sum_{k=0}^{5} l_k.$$

Then the interpolation is computed as follows:

$$L_s(t) = \sum_k \beta_k^s N_s(t-k), \ L_s(k) = \delta_{k,0}, \ k \in \mathbf{Z}. \tag{5.60}$$

By the sequence β_k^s satisfying Equation (5.60), we have

$$f_0(t) = \sum_l c_{0,l} N_s(t-l), \ c_{0,l} = \sum_l f(l) \beta_{k-l}^s, \ l \in \mathbf{Z}, \tag{5.61}$$

and

$$c_{0,k}^R = \sum_l c_{0,l} K_{k-l}^R, \ c_{0,k}^I = \sum_l c_{0,l} K_{k-l}^I, \ l \in \mathbf{Z}, \tag{5.62}$$

where $f_0(t)$ is the approximate input signal. Finally, we obtain the interpolation as follows:

$$f_0(t) = \sum_k c_{0,k}^R N_R(t-k) + \sum_k c_{0,k}^I N_I(t-k), \ k \in \mathbf{Z}. \tag{5.63}$$

Comparing Equations 5.63 and 5.51, it is clear that both $\sum_k c_{0,k}^R N_R(t-k)$ and $\sum_k c_{0,k}^I N_I(t-k)$ terms of Equation 5.63 become the standard forms expressed as the Equation (5.51).

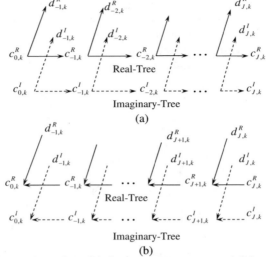

Fig. 5.22. Dual-tree algorithm: (a) decomposition tree, and (b) reconstruction tree

B. Coherent Dual-tree Algorithm

The coherent dual-tree algorithm can be shown as in Figure 5.22. In this algorithm, the real sequences $\{c_{0,k}^R\}$ and the imaginary sequences $\{c_{0,k}^I\}$ are first calculated from $f_0(t)$ by the interpolation expressed as Equations 5.61 and 5.62. Then following the decomposition tree shown in Figure 5.22a, they are decomposed ordinarily by Equations 5.64 and 5.65:

$$c_{j-1,k}^R = \sum_l a_{l-2k}^R c_{j,l}^R, \quad d_{j-1,k}^R = \sum_l b_{l-2k}^R c_{j,l}^R, l \in \mathbf{Z}, \tag{5.64}$$

$$c_{j-1,k}^I = \sum_l a_{l-2k}^I c_{j,l}^I, \quad d_{j-1,k}^I = \sum_l b_{l-2k}^I c_{j,l}^I, \quad l \in \mathbf{Z}. \tag{5.65}$$

The reconstruction tree shown in Figure 5.22b can be applied. The inverse transformation can be calculated by the following equations:

$$c_{j,k}^R = \sum_l (p_{k-2l}^R c_{j-1,l}^R + q_{k-2l}^R d_{j-1,l}^R), \quad l \in \mathbf{Z}, \tag{5.66}$$

$$c_{j,k}^I = \sum_l (p_{k-2l}^I c_{j-1,l}^I + q_{k-2l}^I d_{j-1,l}^I), \quad l \in \mathbf{Z}. \tag{5.67}$$

By Equation 5.45, we have

$$\begin{aligned}\langle \psi_{j,k}^R, \psi_{j,k}^I \rangle &= 0, \\ ||\psi_{j,k}^R|| &= ||\psi_{j,k}^I|| = 1.\end{aligned} \tag{5.68}$$

The norm of the synthetic wavelet can be computed as follows:

$$||d_{j,k}^R \psi_{j,k}^R + d_{j,k}^I \psi_{j,k}^I|| = \sqrt{(d_{j,k}^R)^2 + (d_{j,k}^I)^2}. \tag{5.69}$$

As shown above, our coherent dual-tree algorithm is very simple and it is not necessary to provide the delay of one tree's filter, which is one sample offset from another tree's filter in level -1. Therefore, complex analysis can be carried out coherently all analysis levels.

5.3.4 2-D Complex Discrete Wavelet Transforms

A. Extending the 1-D CDWT to 2-D CDWT

We summarize how the 1-D DWT is extended to the 2-D DWT. First, each row of the input image is subjected to a level -1 wavelet decomposition. Then each column of these results is subjected to a level -1 wavelet decomposition. In each decomposition, the data is simply decomposed into a high frequency component (H) and a low frequency component (L). Therefore, in level -1 decomposition, the input image is divided into HH, HL, LH, and LL components. We denote high frequency in the row direction and low frequency in the column direction as HL and so on. The same decomposition is continued recursively for the LL component.

Following the above procedure, we extend 1-D CDWT to 2-D CDWT. As shown in Figure 5.23a, each row of the input image is first subjected to 1-D RI-spline wavelet decomposition; one is a real decomposition that uses the real component of the RI-spline wavelet and the other is an imaginary decomposition that uses the imaginary component of the RI-spline wavelet. Then each column of these results is also subjected to a 1-D RI-spline wavelet decomposition. In this way, we obtain level -1 decomposition results. When level -1 decomposition is finished, we obtain four times as many results as with the ordinary 2-D DWT decomposition. That is, the 2-D CDWT has four decomposition types; RR, RI, IR, and II are shown in Figure 5.23a. We denote a real decomposition in the row direction and an imaginary decomposition in the column direction as RI and so on. Note that each of these decompositions has HH, HL, LH, and LL components. Furthermore, for the LL component, the same decomposition by which the LL component has been calculated is continued recursively as shown in Figure 5.23c.

The two dimensional RI-spline wavelet functions of RR, RI, IR, and II can be expressed as follows using the 1-D wavelet functions: $\psi^R(t)$ and $\psi^I(t)$ [5],

$$
\begin{aligned}
\psi^{RR}(x,y) &= \psi^R(x)\psi^R(y), \\
\psi^{RI}(x,y) &= \psi^R(x)\psi^I(y), \\
\psi^{IR}(x,y) &= \psi^I(x)\psi^R(y), \\
\psi^{II}(x,y) &= \psi^I(x)\psi^I(y).
\end{aligned} \tag{5.70}
$$

Figure 5.24 shows these 2-D wavelet functions, where Figure 5.24a shows wavelet function ψ^{RR}, Figure 5.24b the wavelet function ψ^{IR}, Figure 5.24c the

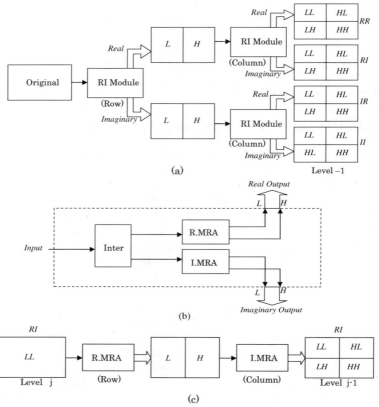

Fig. 5.23. The 2-D CDWT implementation and definition: (a) block diagram of level -1, (b) RI module, (c) Block diagram from level j to level $j - 1$ when $j < -1$

wavelet function ψ^{RI} and Figure 5.24d the wavelet function ψ^{II}. Comparing Figures 5.24a, b, c and d, it is clear that the wave shapes of ψ^{RR}, ψ^{RI}, ψ^{IR} and ψ^{II} are different, so different information can be extracted by using them.

Moreover, based on Equations 5.68 and 5.70, the norm of the 2-D RI-spline wavelet function ψ_j^{RR} in a point (k_x, k_y) of level j, $||\psi_j^{RR}(x - k_x, y - k_y)||$, which is abbreviated to $||\psi_{j,k_x,k_y}^{RR}||$ hereafter, can be expressed as follows:

$$||\psi_{j,k_x,k_y}^{RR}|| = ||\psi_{j,k_x}^{R}\psi_{j,k_y}^{R}||^2$$
$$= ||\psi_{j,k_x}^{R}||^2||\psi_{j,k_y}^{R}||^2 \qquad (5.71)$$
$$= 1.$$

The same is true for the other wavelet functions ψ_j^{RI}, ψ_j^{IR} and ψ_j^{II}. Furthermore, the inner product of ψ_{j,k_x,k_y}^{RR}, ψ_{j,k_x,k_y}^{RI}, ψ_{j,k_x,k_y}^{IR}, and ψ_{j,k_x,k_y}^{II} is zero since $< \psi_{j,k}^{R}, \psi_{j,k}^{I} >= 0$. This means that the 2-D wavelet functions shown in Equation 5.70 are orthogonal to each other. The 2-D synthetic wavelet coefficients

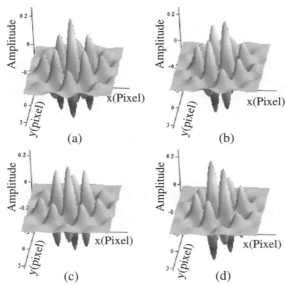

Fig. 5.24. Example of two dimension RI-spline wavelets: (a) wavelet function ψ^{RR}, (b) wavelet function ψ^{IR}, (c) wavelet function ψ^{RI}, and (d) wavelet function ψ^{II}

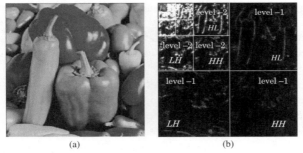

Fig. 5.25. Norm obtained by 2-D CDWT using an m=4,3 RI-spline wavelet from level -1 to level -4: (a) 256×256 Pepper image, (b) 2-D TI coefficients obtained by using the RI-spline wavelet

$|d_{j,k_x,k_y}|$ in HH of RR, RI, IR, and II that were obtained in level j (k_x, k_y) by 2-D CDWT using the 2-D RI-Spline can be defined as follows:

$$|d_{j,k_x,k_y}| = \sqrt{(d_{j,k_x,k_y}^{RR})^2 + (d_{j,k_x,k_y}^{RI})^2 + (d_{j,k_x,k_y}^{IR})^2 + (d_{j,k_x,k_y}^{II})^2}. \qquad (5.72)$$

In the same way as in the 1-D case, the 2-D synthetic wavelet coefficients $|d_{k_x,k_y}^j|$ become the norm. Thus they can be treated as translation invariant features, because they are insensitive to phase. The same results can be ob-

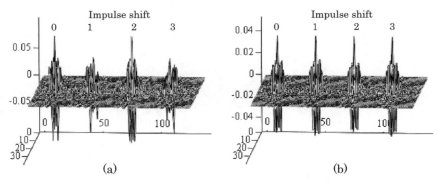

Fig. 5.26. Impulse responses of the $m = 4$ spline wavelet and $m = 4, 3$ RI-spline wavelet on level-2: (a) impulse responses of HH in level -2 using the $m = 4$ spline wavelet, and (b) Impulse responses of HH in level -2 using the $m = 4, 3$ RI-spline wavelet

tained in the case of the LH and HL. Hereafter we call $|d_{j,k_x,k_y}|$ translation invariant (TI) coefficients.

Figure 5.25b shows an example of the 2-D TI coefficients from level -1 to level -4 that were obtained by the 2-D CDWT applied to the original image shown in Figure 5.25a. As shown in Figure 5.25, it is clear that the 2-D synthetic wavelet in LH, HH, HL carries the intrinsic information of each local point.

In order to demonstrate the translation invariance an experiment of the 2-D CDWT was performed. Figure 5.26a shows the impulse response of the 2-D $m = 4$ spline wavelet, and Figure 5.26b shows the impulse response of the 2-D $m = 4, 3$ RI-spline wavelet. Here, "impulse response" means the following. The input images have an "impulse" when only one pixel is 1, and the others are 0. Then the horizontal position of impulse shifts one by one. These "impulse" input images are subject to 2-D DWT using an $m = 4$ spline wavelet and 2-D CDWT using an $m = 4, 3$ RI-spline wavelet. After being subjected to 2-D DWT and 2-D CDWT, only the coefficients (in the case of 2-D CDWT, the coefficients mean the TI coefficients) of HH in level -2 are retained, and other coefficients are rewritten to be 0. These coefficients are used for reconstruction by the inverse transform. "Impulse response" means that these reconstructed images are overwritten. If the shapes of these "impulse responses" have the same independence of the position of the impulse, the wavelet transform used to make the "impulse response" can be considered as being translation invariant. Comparing Figures 5.26a and 5.26b, the "impulse response" of the 2-D CDWT has uniform shape while that of the 2-D CWT does not.

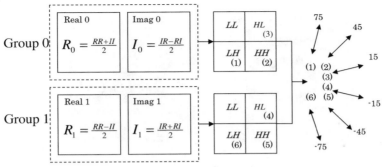

Fig. 5.27. Calculations of the directional selection

B. Implementation of Directional Selection by 2-D CDWT

As shown in Figure 5.27, direction selectivity can be implemented by calculating the sum or difference between the wavelet coefficients of the four kinds of RR, RI, IR, and II that were obtained in 2-D CDWT. We here take the direction 45° of Figure 5.27 as an example and show the details of the calculation method.

If the wavelet coefficients of Equation 5.70 are assumed to be $d^{RR}_{j,kx,ky}$, $d^{RI}_{j,kx,ky}$, $d^{IR}_{j,kx,ky}$ and $d^{II}_{j,kx,ky}$, the calculation in a direction of 45° can be carried out by following the calculation method Real0 and Imag0 shown in Figure 5.27,

$$d^{R0}_{j,kx,ky} = \frac{d^{RR}_{j,kx,ky} + d^{II}_{j,kx,ky}}{2} \text{ (Real)}, \tag{5.73}$$

$$d^{I0}_{j,kx,ky} = \frac{d^{RI}_{j,kx,ky} - d^{IR}_{j,kx,ky}}{2} \text{ (Imaginary)}. \tag{5.74}$$

The waveform of the 45° direction shown in Figure 5.27 will be extracted alternatively by making $d^{R0}_{j,kx,ky}$, $d^{I0}_{j,kx,ky}$ as real and imaginary components of complex wavelet coefficients. Furthermore, directions 75° and form 15° − -75° degrees shown in Figure 5.27 are calculated similarly.

Furthermore, the image of the circle shown in Figure 5.28a was used to test the actual direction selectivity of our approach. The analysis results obtained by conventional 2-D DWT using an $m = 4$ spline wavelet (real type) are shown in Figure 5.28b and results obtained by 2-D CDWT using an RI-spline (complex type) are shown in Figure 5.29. By comparing Figures 5.28b and 5.29, it is clear that in the case of conventional 2-D DWT, only three-direction selectivity was acquired and especially the 45° direction cannot be separated. On the other hand, in the case of the 2-D CDWT, it turns out that six directions were extracted in a stable manner.

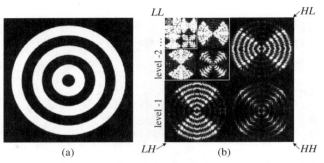

Fig. 5.28. Circle image (256×256) and directional selection obtained by 2-D CDWT using a RI-spline wavelet: (a) circle image and (b) analysis result obtained by 2-D DWT

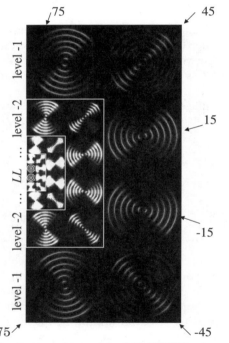

Fig. 5.29. Directional selection obtained by 2-D CDWT using an RI-spline wavelet

5.4 Applications in Signal and Image Processing

5.4.1 Fractal Analysis Using the Fast Continuous Wavelet Transform

A. Fractal Analysis

The fractal property indicates the self-similarity of the shape structure or phenomenon. Self-similarity means that the shape, structure and phenomenon are not changed even if their scales are expanded or reduced. Though the strict self-similarity is recognized in only regular fractal figures, the shape, structure and phenomenon that have self-similarity exist in a scale range in nature (e.g. the shape of clouds and coastlines, and the structure of turbulent flow).

Most of the time series signals indicate a continuous structure that has no frequency component with a remarkable power. Generally, the power of a lower frequency component of this kind of signal is larger than that of a higher one. Furthermore most of these kinds of signals can be approximated by using the function of $E(f) \propto f^{-\alpha}$ in some ranges. For example, turbulence is one of them. If the power spectrum of the time series can be approximated by using $f^{-\alpha}$, the value of α is an exponent representing the self-similarity of the signal. The relation between the exponent α and the fractal dimension D can be shown as in Equation 5.75 in the range of $1.0 < \alpha < 3.0$ [21, 22].

$$D_f = \frac{(5 - \alpha)}{2}. \tag{5.75}$$

Here, the condition where the signal has a fractal property is judged by whether the power spectrum of signal can be approximated by using $f^{-\alpha}$, and the fractal dimension obtained from the power spectrum using Equation 5.75 is defined as D_f.

The dimension that we generally consider indicates the free degree of space. For example, a segment is one-dimensional, and a square is two-dimensional. Generally, the figure that has a fractal property is of very complex shape, whose complexity is quantified by a non-integer dimension. Thus, the dimension expanded to the set of non-integral values is called the fractal dimension or generalized dimension. One such dimension is the similarity dimension and, for example, the well-known Koch curve having a similarity ratio of $1/3$ and four similarity figures, and its similarity dimension is non-integer 1.2618. Therefore, if fractal analysis is used, the complexity of shape, structure and phenomenon that generally cannot be evaluated quantitatively can be evaluated by using a non-integer dimension.

B. Computation Method of Fractal Dimension

A fractal dimension can be obtained by changing the scale value. The CWT expands an unsteady signal into the time-frequency plane and can examine the time change of its frequency. The process of downsampling in CWT is the same as that of the fractal analysis of unsteady signals. Consequently, both analysis methods have a connection with each other. Therefore, we show the following scale method using CWT based on this relation. (1) The scale degree is represented by $a_m = 2^{i/M} 2^j$. The average of absolute values of the difference from the wavelet coefficients is calculated by $E(|w(a_{m+1}, b) - w(a_m, b)|)$ for each scale. They are plotted in a bi-logarithmic graph, and the gradient of the line approximated by the least squares method is P. This operation is similar to the calculation of the gradient of the power spectrum from $w(a, b)$ at each time,

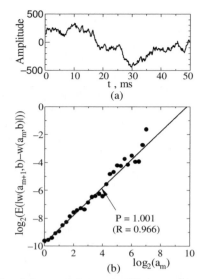

Fig. 5.30. An example of the model signal of Brownian motion with dimension 1.5 and its characteristic value P analyzed using WSM: (a) model signal with dimension 1.5 and (b) result analyzed by using WSM

$$P = \frac{\log_2 E(|w(a_{m+1}, b) - w(a_m, b)|)}{\log_2 a_m}. \tag{5.76}$$

(2) The relationship between P and D is obtained by using the model signal where the fractal dimension D is already known, and the fractal dimension D_w is determined from P. This method is called the wavelet scale method (WSM).

C. Determination of the Fractal Dimension

Brownian motion is random and fractal, and its power spectrum can be approximated by using the function of $E(f) \propto f^{-\alpha}$ [22]. Therefore, The Brownian motion that is a useful model signal to determine the fractal dimension D_w was considered. Figure 5.30a shows an example of the model signal of fractional Brownian motion having the fractal dimension $D = 1.5$, and Figure 5.30b is its characteristic quantity obtained by Equation 5.76. The octave number of the analysis is seven, and each octave is divided into four ($M = 4$ voices) in CWT. The data length is 512 points, and the sampling frequency is 10 kHz. As shown in Figure 5.30, $\log_2 E(|W(a_{m+1}, b) - W(a_m, b)|)$ shows a good linearity to $\log_2 a_m$, and it is recognized that the high correlation value $R = 0.966$ was obtained.

Next, model signals with $D = 1.1 \sim 1.9$ are made, and the relation between P and D is determined as shown in Figure 5.31. P and the correlation coefficients R in Figure 5.31 are the average value of 10 sets of model signals,

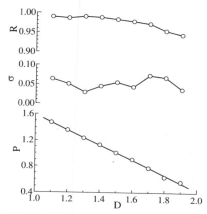

Fig. 5.31. Values of R, σ and the relation between P and D

and the variance of P is expressed with the standard deviation σ. As shown in Figure 5.31, P decreases as the fractal dimension D increases from 1.1 to 1.9 and the variance of P is about 0.05. The minimum value of the correlation coefficient R is about 0.950, although R shows the tendency of a little decrease with an increase in fractal dimension, that is, high correlation was obtained. Furthermore, the fractal dimension D of the model signals is plotted versus P and a straight line obtained by the least squares method as follows,

$$D_w = -0.853 \times P + 2.360. \tag{5.77}$$

A high correlation value of 0.989 between D_w and P was obtained. That is, the fractal dimension of a signal can be evaluated using D_w obtained above.

Generally, the fractal dimension was calculated using long data and only the mean fractal dimension was obtained. Oppositely, WSM can calculate the fractal dimension in each time theoretically and find out the time change of the fractal dimension of an unsteady signal, since WSM uses the wavelet transform that can analyze both time and frequency at the same time. However, the fractal dimension obtained may produce small variances and calculation accuracy may become lower since there are fewer data in each time. Therefore, it is necessary for the average interval to be set up to increase the number of data in order to improve calculation accuracy.

Figure 5.32a shows the D_w that was obtained by WSM and Figure 5.32b shows its standard deviation, where the average data numbers are 16, 32, 64, 128, 256, respectively, and the model data used is Brownian motion with a fractal dimension of 1.5. The wavelet transform was carried out under the two conditions of $M = 4$ and $M = 8$. As shown in Figure 5.32a, the mean D_w mostly shows a fixed value even if the average data number and M are changed. On the other hand, the standard deviation shown in Figure 5.32b tends to become large as the average data number becomes small. The same results can be obtained in the case of $D = 1.2, 1.8$, respectively. In addition,

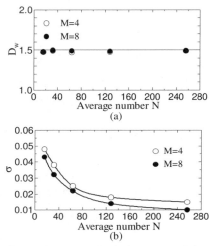

Fig. 5.32. Averaged D_w and its standard deviation: (a) averaged dimension D_w and (b) standard deviation of D_w

the standard deviation becomes small when the number M increases in the same average data number and the calculation time becomes large. This is because the number of average data increase. Therefore, the voices $M = 4$ were chosen for computation efficiency. In this case, data number of 64 or more is desirable in order to obtain the variance of D_w below 3%, and then the time change of the fractal dimension by D_w can be evaluated with good accuracy.

D. Fractal Analysis of the Tumbling Flow in a Spark Ignition Engine

The tumbling flow is often seen in high-speed spark ignition engines with a pentroof-type combustion chamber and four valves (two valves for both intake and exhaust), and keeps the kinetic energy that is introduced by gas flow in the intake stroke and breaks down in the latter stage of the compression stroke. Therefore, it is considered that the tumbling flow is effective for promoting combustion because it is converted into many smaller eddies before the top dead center (TDC) and the turbulence intensity increases. We show here the change of the eddies' structure before and after the tumbling flow breaks down.

The gas flow velocity in the axial direction at a position of 5 mm from the cylinder center was measured with an LDV under the condition of motoring and an engine speed of $n = 771$ rpm. The engine for the experiment had four valves, and the bore and stroke are 65 mm and 66 mm, respectively. The examples of the fluctuation velocity of the tumbling flow (frequency components of more than 100Hz) in $220°$–$450°$ (TDC is $360°$) is shown in Figure 5.33 versus crank angle. These power spectra, which correspond to a data length of

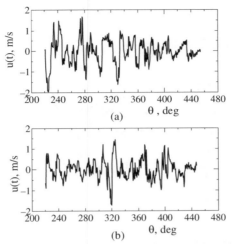

Fig. 5.33. Examples of the fluctuation velocity $u(t)$: (a) velocity $u(t)$ in the condition $\varepsilon = 3.3$ $n = 771$ rpm and (b) in the condition $\varepsilon = 5.5$ $n = 771$ rpm.

51.2 ms (512 samples) from 220° to 450°, where Figure 5.33a is an example in compression ratio $\varepsilon = 3.3$, and Figure 5.33b is an example in $\varepsilon = 5.5$.

In the case of $\varepsilon = 3.3$ (Figure 5.33a) where the tumbling flow does not break down, the $u(t)$ becomes smaller with increasing crank angle and its $D_w = 1.672$ can be obtained using the data from 220° to 450°. Oppositely, in the case of $\varepsilon = 5.5$ (Figure 5.33b), $u(t)$ first becomes large by the tumbling flow breaks down near the 320° crank angle and then becomes small with increasing crank angle. $D_w = 1.710$ can be obtained using the data from 220° to 450°.

Furthermore, the change in fractal dimension D_w is calculated by the WSM, and results are shown in Figure 5.34, where in order to reduce the variance of D_w within 3%, the average length of 100 points (10 ms) has been adopted. Figure 5.34a is obtained from the fluctuation velocity when $\varepsilon = 3.3$, $n = 771$ rpm which is shown in Figure 5.33a, and 5.34b is obtained when $\varepsilon = 3.3$, $n = 771$ rpm which is shown in Figure 5.33.

As shown in Figures 5.34a and 5.34b, in the case of $\varepsilon = 3.3$, small eddies are the strongest near 300° crank angles and $D_w = 1.692$ at first, then they decrease with compression and become $D_w = 1.309$ near TDC. However, in the case of $\varepsilon = 5.5$ as shown in Figure 5.34b, the fractal dimension decreases a little near 320° crank angles because the tumbling flow is broken down and the energy of larger eddies of the fluctuation becomes large as shown in Figure 5.33b. Then, the fractal dimension increases and becomes $D_w = 1.591$ near TDC because the energy of small eddies becomes larger. That is, the eddies that are generated by the tumbling flow which has broken down have a larger scale and transmit the energy to the small eddies in the compression stroke. After TDC, the small eddies in the gas flow here also generated by the

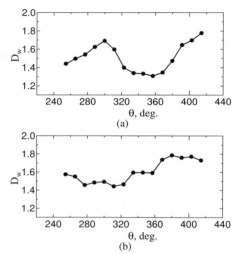

Fig. 5.34. Fractal dimension D_w of the fluctuation velocity $u(t)$: (a) result obtained in the condition $\varepsilon = 3.3$ and (b) in the condition $\varepsilon = 5.5$.

piston motion and the energy in the power spectra increases. Consequently, the fractal dimension increases in both compression ratios of 3.3 and 5.5 as shown in Figures 5.34a and 5.34b. Therefore, it is clearly shown on the above discussion that the proposed fractal dimension D_w is effective for evaluating the change in the structure of the eddies quantitatively.

5.4.2 Knocking Detection Using Wavelet Instantaneous Correlation

A. Analysis of the Knocking Characteristics

In engine control, knocking detection is an important problem and a lot of research has been published on this over many years [23, 24]. The conventional knocking detection method is generally effective at lower engine speeds, where the signal-to-noise ratio (SNR) is high. However, because SNR decreases significantly at high engine speeds, the method has difficultly detecting the knocking precisely. Actually, in the region of high engine speeds, a compromise method that does not detect the knocking and sets up the ignition time retardation beforehand was used, although the method sacrifices engine performance. A detection method that rejects high engine noise at high engine speeds would therefore be desirable in order to obtain the original performance of the engine. In this study, we try using the WIC method to extract knocking signals at high engine speeds.

Knocking experiments were carried out by a bench test. The engine has four cylinders in line and gasoline is injected into the intake pipe. The combustion chamber is of pent-roof type, the bore and stroke of each cylinder

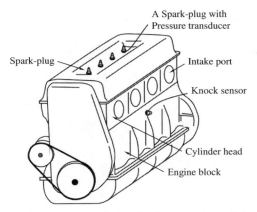

Fig. 5.35. Test engine and attachment position of the sensors

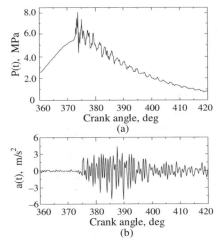

Fig. 5.36. Example of pressure and vibration signals obtained under knocking conditions: (a) pressure signal and (b) vibration signal

is 78.0 mm and 78.4 mm, respectively, with a compression ratio of 9.0. As shown in Figure 5.35, a spark plug with a piezoelectric pressure transducer (KISTLER 6117A) was used in cylinder 4 to measure the pressure history. The engine block vibration was measured with a knock sensor. The sensor was placed on the side of the engine between cylinders 2 and 3. Experiment 1 measured the combustion pressure and block vibration under a full loaded condition, and the engine speed was kept at $n = 3000$, 5000 and 6000 rpm, respectively. Experiment 2 tested under knocking conditions. The engine also operated under the same load and speed conditions, where various degrees of light and heavy knocks were induced by advancing the ignition time.

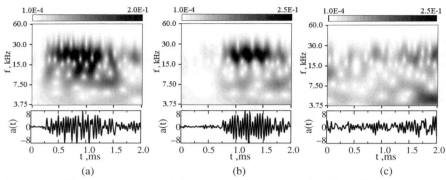

Fig. 5.37. Wavelet transform of the vibration signals obtained by a knocking sensor: (a) analysis result obtained in the condition of heavy knocking, (b) in the condition of light knocking and (c) in the condition of normal combustion

An example of the pressure signal and the block vibration signal measured at $n = 3000$ rpm is shown in Figure 5.36, where 360° corresponds to the top dead center (TDC). As shown in Figure 5.36, when a knock occurs, there is corresponding vibration of the pressure and vibration of the engine block. Figure 5.37 shows an example of the wavelet transform of the vibration of the engine block at $n = 3000$ rpm, where time 0 corresponds to the time of 10° after TDC, and Figure 5.37a is in a state of heavy knocking, Figure 5.37b in a light-knocking state and Figure 5.37c in a normal combustion state. To carry out the CWT, the RI-spline wavelet shown in Sect. 5.2.2 was used as the MW and the high-speed computation method in the frequency domain shown in Section 5.2.3 was used. The vibration signals were normalized as its standard deviation $\sigma=1$ in order to suppress the influence of the amplitude of the signal and to make the characteristic of the frequency clear. The ordinates in Figure 5.37 denote frequency and transverse time. The amplitude of $|w(a,b)|$ is shown as the shade level, the analyzing frequency range chosen was four octaves and each octave was divided into 48 voices.

As shown in Figures 5.37a and b, the pattern centering on about 20 kHz was strongly detected from the vibration by the knocking. This is because the knocking sensor used for this experiment had a large sensitivity to frequency components above 17 kHz. Next, in Figure 5.37c, which represents normal combustion, the pattern centering on about 20 or 40 kHz does not exist. Correspondingly, the pattern centering on about 20 kHz of the vibration signals can be treated as a characteristic pattern of knocking (which consists of two or more frequency components and amplitudes of each frequency component which changes with time).

B. Constructing the RMW Using the Knocking Signal

The characteristic part of the knocking in the vibration signal shown in Figure 5.37b, which was extracted from the neighborhood for 1.1 ms and has

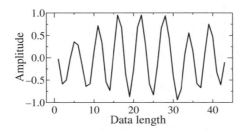

Fig. 5.38. Vibration signal cut out from 1.1ms for 43 points in the Figure 5.37b

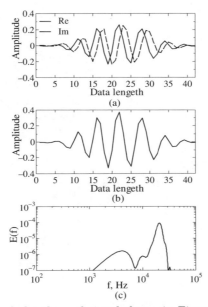

Fig. 5.39. RMWs made by the real signal shown in Figure 5.38 and their power spectrum: (a) complex type RMW, (b) real type RMW, and (c) their power spectrum

a length of 43 samples in the sampling time 0.175 ms, was shown in Figure 5.38. The complex RMW was constituted by using the method shown above and is shown in Figure 5.39b. For comparison, the real RMW was also constructed and is shown in Figure 5.39a, and its power spectrums are shown in Figure 5.39c. As is shown in Figure 5.39c, RMW has a big peak centered at about 20 kHz and small peaks with lower frequency in the frequency domain, that is, it can be observed that it has two or more feature components.

C. Detecting Knocking Signals by WIC

The values of $R(b)$ are calculated from the vibration at engine speeds of $n = 3000$, 5000, and 6000 rpm, respectively, and the results are shown in Figure 5.40, where time $t = 0$ denotes 10° after TDC. Figure 5.40I shows the

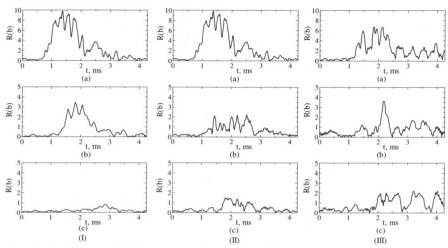

Fig. 5.40. Values of $R(b)$ obtained from wavelet instantaneous correlation, where (I) shows result obtained in the condition of $n = 3000$ rpm, (II) $n = 5000$ rpm, (III) $n = 6000$ rpm: (a) heavy knocking (b) light knocking and (c) normal combustion

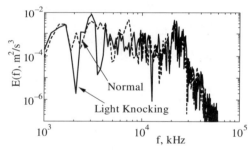

Fig. 5.41. Power spectrums of vibration in the case of light knocking and normal combustion

results obtained at 3000 rpm, Figure 5.40II at 5000 rpm, and Figure 5.40III at 6000 rpm. Figure 5.40a denotes the case of strong knocking, Figure 5.40b light knocking, and Figure 5.40c normal combustion. As shown in Figure 5.40, the amplitude of $R(b)$ changes with the knocking strength in the same engine speed, that is, the generating time of the knocking and the strange of knocking can be evaluated simultaneously by using the amplitude of $R(b)$.

In addition, in normal combustion, it is observed that the value of $R(b)$ increases as the engine speed increases. This is because the amplitude of the noise becomes large as the engine speed becomes high. By comparing the value of $R(b)$ between light knocking and normal combustion at 5000 and 6000 rpm, the difference in the light knocking and normal combustion is clearly distinguishable. Moreover, the power spectrums of the light knocking and normal combustion at 6000 rpm shown in b and c of Figure 5.40III were obtained and are shown in Figure 5.41. As shown in Figure 5.41, the difference between light knocking and normal combustion from a power spectrum was hardly observed, so light knocking can not be distinguished from normal combustion.

5.4.3 De-noising by Complex Discrete Wavelet Transforms

It is well-known that the wavelet shrinkage proposed by Donoho and Johnstone, which uses the DWT, is a simple but very effective de-noising method [25]. However, it has been pointed out that de-noising by wavelet shrinkage sometimes exhibits visual artifacts, such as the Gibbs phenomena, near edges. This is because ordinary DWT lacks translation invariance [6]. In order to overcome such artifacts, Coifman and Donoho [26] proposed a translation invariant (TI) de-noising method. In their TI de-noising, they averaged out the artifacts, which they called "cycle spinning", so the de-noised results became translation invariant. That is, they used a range of shifts of the input data, then de-noised by wavelet shrinkage, and averaged the results. Romberg *et al.* [27] extended this method and applied TI de-noising to image de-noising. Cohen *et al.* [28] proposed another TI de-noising method using shift-invariant wavelet packet decomposition. The common drawback of all these methods lies in their computational time, that is, in exchange for achieving translation invariant de-noising, all of these methods increase the computational time considerably.

In this Sect., we show a different approach to creating TI de-noising, namely that we apply the translation invariant CDWT using an RI-spline wavelet to TI de-noising. Furthermore, de-noising experiments with the ECG data were carried out.

A. Ordinary Wavelet Shrinkage

Wavelet shrinkage is a well-known de-noising method that uses the wavelet decomposition and reconstruction enabled by orthonormal DWT [25]. In order to remove noise, Donoho and Johnstone proposed that only the wavelet coefficients undertaking a *soft thresholding* operation, which is expressed as Equation 5.78, should be used for the signal reconstruction,

$$|\hat{d}_{j,k}| = \begin{cases} |d_{j,k}| - \lambda & |d_{j,k}| > \lambda \\ 0 & |d_{j,k}| \le \lambda \end{cases}. \tag{5.78}$$

Donoho and Johnstone also proposed that the universal threshold λ, which is decided by Equation 5.78, should be used in every decomposition level,

$$\lambda = \sigma \sqrt{2 \log_e(N)}, \tag{5.79}$$

where N denotes the sample number and σ the standard deviation of the white noise to be removed. Notice that the reason that the universal threshold λ is used in every decomposition level corresponds to the fact that the

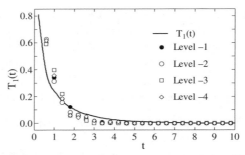

Fig. 5.42. Distribution obtained by $(d_{j,k})^2$

spectrum of white noise is flat when using orthonormal DWT. In real situations, however, the σ of the noise to be removed is unknown. Thus Donoho and Johnstone proposed practical methods to estimate this σ using the following equation:

$$\sigma = \frac{\text{median}||d_{J,k}| - \text{median}(|d_{J,k}|)|}{0.6745}, \qquad (5.80)$$

where $J = -1$.

B. Wavelet Shrinkage Using an RI-Spline Wavelet

Here we augment wavelet shrinkage using the translation invariant RI-spline wavelet. We first define $|d_{j,k}|$ as follows:

$$|d_{j,k}| = \sqrt{(d_{j,k}^R)^2 + (d_{j,k}^I)^2}. \qquad (5.81)$$

Based on Equation 5.69, we can call $|d_{j,k}|$ calculated by using Equation 5.81 the norm. Hereafter we call $|d_{j,k}|$ translation invariant (TI) coefficients. However, we cannot apply the inverse wavelet transform operation to the norm expressed by Equation 5.81. Thus after the norm has undergone the thresholding operation expressed as Equation 5.82, the real and imaginary components should be subjected to the following operations

$$\hat{d}_{j,k}^R = d_{j,k}^R \frac{|\hat{d}_{j,k}|}{|d_{j,k}|}, \quad \hat{d}_{j,k}^I = d_{j,k}^I \frac{|\hat{d}_{j,k}|}{|d_{j,k}|}. \qquad (5.82)$$

$\hat{d}_{j,k}^R$ and $\hat{d}_{j,k}^I$ are used for the inverse wavelet transform shown in Figure 5.22(b) instead of $d_{j,k}^R$ and $d_{j,k}^I$.

As is shown above, ordinary wavelet shrinkage uses orthonormal DWT, for which Equation 5.79 is optimized. As the RI-spline wavelet uses a pair of bi-orthonormal wavelets, we cannot use the threshold λ decided by Equation 5.79.

As Equation 5.79 is decided statistically, we also use a statistical method to determine the threshold, so that for RI-spline wavelet the threshold should

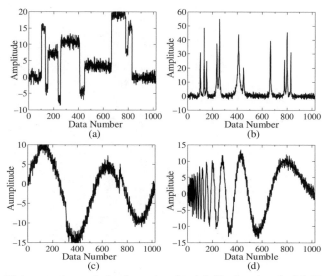

Fig. 5.43. Noisy versions of the four signals: (a) blocks signal, (b) Bumps signal, (c) HeaviSine signal, (d) Doppler signal. White noise $N(0, 1)$ has been added in each case (SNR = 17 dB)

have the equivalent statistical meaning as the threshold λ decided by Equation 5.79. When random variables X_1 and X_2 are independent of each other, $(X_1)^2 + (X_2)^2$ follows the chi-square distribution $T_2(t)$. However $d_{j,k}^R$ and $d_{j,k}^I$ shown in Equation 5.81 are not exactly independent. Figure 5.42 shows the distribution of $(d_{j,k})^2$ obtained by Equation 5.81 when the signal is Gaussian white noise with $\sigma = 1$, $\mu = 0$. The solid line denotes the theoretical distribution $T_1(t)$ $(t = x^2)$, the marks show the distribution of $(d_{j,k})^2$ obtained in different levels. Notice that j corresponds to the level. As is shown in Figure 5.42, the distribution of $(d_{j,k})^2$ is approximated by $T_1(t)$ in every level. Thus we also use the same threshold for every level. The probability that $t \leq 10.83$ is 99.9% can be obtained from the distribution $T_1(t)$. In order to have the equivalent statistical meaning as Equation 5.79, one assumes $\lambda = \sqrt{10.83} = \sqrt{a \log_e(N)}$ when $\sigma = 1$ and obtains $a = 1.56$ approximately. Finally, the threshold value λ in the case of the 1-D CDWT using the RI-spline wavelet is determined as:

$$\lambda = \sigma \sqrt{1.56 \log_e(N)}. \tag{5.83}$$

C. Experimental Results Obtained by Using Model Signals

Following the experiments by Coifman [26], we use four types of model signals: Blocks, Bumps, HeaviSine and Doppler for experiments. Gaussian white noise $N(0, 1)$ has been added to these four model signals with SNR = 17 dB (SNR is the ratio of the signal power and the noise power, and is shown in dB) to create

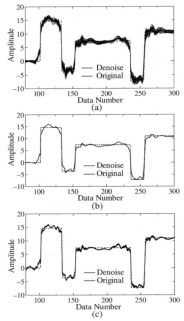

Fig. 5.44. The original signal and example of the de-noised signals obtained by D8, TI-D8 and the RI-spline wavelet when shifts are from 0 to 15 samples: (a) de-noising result with D8, (b) de-noising result with D8-TI and (c) de-noising result with RI-SP

the noisy signals shown in Figure 5.43, where the noise SNR is same as in [26]. Figure 5.44 shows an example of the de-noised results, where Figure 5.44a shows the overwritten results obtained by Daubechies 8 (D8) with 16 sample shifts, Figure 5.44b shows the averaged result of 16 sample shifts shown in Figure 5.43a using the TI de-noising method [26], which uses D8 (TI-D8), and Figure 5.44c shows the overwritten results with 16 sample shifts using the $m = 4, 3$ RI-spline wavelet (RI-SP). Samples are shifted one by one from 0 to 15 shift. For determining the threshold value λ, Equation 5.79 was used for the Daubechies wavelet and Equation 5.83 was used for the RI-SP wavelet. From Figure 5.44a, it is apparent that the results de-noised by D8 vary with sample shifts. However, from Figurae 5.44c, this phenomenon cannot be observed in de-noised results by RI-SP, and the results are comparable to those by TI-D8. In addition, in both Figures 5.44b and 5.44c, the Gibbs phenomenon around the corner has been suppressed. Next, the root mean squared errors (RMSE) between the de-noised signals and the original signals are calculated, which are shown in Figure 5.45. Figure 5.45a shows results obtained in the case of Blocks, and Figure 5.45b obtained in the case of Bumps, Figure 5.45c obtained results in the case of HeaviSine and 5.45d obtained in the case of Doppler. From the Figure 5.45, it is clear that in the RMSE obtained by D8, a large

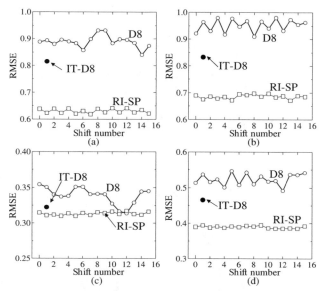

Fig. 5.45. RMSE obtained by D8, TI-D8 and the RI-spline wavelet: (a) results obtained by using Blocks signal, (b) Bumps signal, (c) HeaviSine signal, and (d) Doppler signal

oscillation with sample shifts is observed. The same phenomenon can also be observed in the case of another orthogonal mother wavelet, for example, the Symlet wavelet and so on. This is because conventional DWT is not translation invariant. In contrast, the RMSE using the RI-SP wavelet does not vary in dependence on sample shifts.

In addition, the RMSE obtained using the RI-SP wavelet is smaller than that using TI-D8, although TI-D8 increases the computational time greatly. Similar results were also obtained in other noise SNR conditions. These results clearly show that wavelet shrinkage using the translation invariant RI-spline wavelet can achieve TI de-noising and shows a better performance for de-noising than conventional wavelet shrinkage using DWT.

D. ECG De-noising

It is well-known that an electrocardiogram (ECG) is useful for diagnosing cardiac diseases in children and adults. However, clinically obtained ECG signals are often contaminated with a lot of noise, especially in the case of fetal ECG. In order to remove the noise from ECG signals, many methods have been proposed and especially those using wavelet shrinkage have attracted attention [29, 30]. Examples of removing white noise from an electrocardiogram (ECG) are shown in Figure 5.46. Figure 5.46a shows the signal of the electrocardiogram of an adult that contains white noise with SNR = 12 dB, Figure 5.46b shows the de-noising result obtained by the $m = 4, 3$ RI-spline

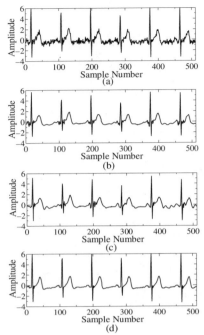

Fig. 5.46. Results of ECG de-noising: (a) ECG signal with white noise SNR = 12 dB (b) de-noising result obtained by using the RI-spline wavelet (c) that obtained by using Daubechies 8 wavelet, (d) the original ECG data

wavelet (RI-SP), Figure 5.46c shows the result obtained by the Daubechies 8 wavelet (D8) and Figure 5.46d shows the original data. By comparing Figures. 5.46b, c and d, it is clear that in the de-noised result by RI-SP shown in Figure 5.46b, we observe less vibration in the waveform than that by D8 (Figure 5.46c).

For quantitative estimation, we calculate the distortion that is the square of the differences between the original signal $f(t)$(Figure 5.46d) and the reconstructed signal $y(t)$. Figure 5.47 shows these results using RI-SP and D8 wavelets with varying SNRs. As shown in Figure 5.47, less distortion is obtained by the RI-SP wavelet than that obtained by the D8 wavelet, in every SNR. For example, when SNR = 12 dB, about 2.5 dB of distortion can be improved by using the RI-SP wavelet.

Furthermore, Figure 5.48 shows the de-noised result of a fetal ECG (38th week of pregnancy). Figure 5.48a shows the original fetal ECG[29], Figure 5.48b shows the de-noised result obtained by DWT using D8 and Figure 5.48c shows the de-noised result obtained by CDWT using the RI-SP. As shown in Figure 5.48a, the fetal ECG includes a lot of noise, and we cannot extract characteristics such as fetal QRS, or P and T waves, although these characteristic waves are important for diagnosing cardiac diseases. In the de-

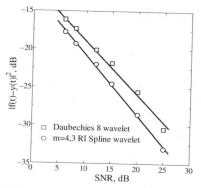

Fig. 5.47. Distortion of the ECG wave after de-noising: (a) original signal of the fetal ECG, (b) de-noising result by DWT using Daubechies 8 and (c) de-noising result by CDWT using the $m = 4, 3$ RI-spline

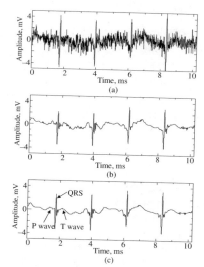

Fig. 5.48. Example of the fetal ECG wave after de-noising

noised result by RI-SP shown in Figure 5.48c, we can look for the fetal QRS, or P and T waves clearly and also observe less vibration in the waveform than that obtained by D8 (Figure 5.48b). These experiments above show that translation invariant de-noising using a translation invariant RI-spline wavelet is effective for real ECG data.

5.4.4 Image Processing and Direction Selection

A. Medical Image De-noising

Following the method shown in Section 5.4.3, when using 2-D CDWT the thresholding operation (soft–thresholding) is carried out by Equation 5.78 using $|d_{j,k_x,k_y}|$ instead of $|d_{j,k}|$. After the TI coefficients have undergone the thresholding operation, each coefficient of RR, RI, IR and II should be subjected to the following operations.

$$
\begin{aligned}
\hat{d}_{j,k_x,k_y}^{RR} &= d_{j,k_x,k_y}^{RR} \frac{|\hat{d}_{j,k_x,k_y}|}{|d_{j,k_x,k_y}|}, \\
\hat{d}_{j,k_x,k_y}^{RI} &= d_{j,k_x,k_y}^{RI} \frac{|\hat{d}_{j,k_x,k_y}|}{|d_{j,k_x,k_y}|}, \\
\hat{d}_{j,k_x,k_y}^{IR} &= d_{j,k_x,k_y}^{IR} \frac{|\hat{d}_{j,k_x,k_y}|}{|d_{j,k_x,k_y}|}, \\
\hat{d}_{j,k_x,k_y}^{II} &= d_{j,k_x,k_y}^{II} \frac{|\hat{d}_{j,k_x,k_y}|}{|d_{j,k_x,k_y}|}.
\end{aligned}
\tag{5.84}
$$

These $\hat{d}_{j,k_x,k_y}^{RR}$, $\hat{d}_{j,k_x,k_y}^{RI}$, $\hat{d}_{j,k_x,k_y}^{IR}$ and $\hat{d}_{j,k_x,k_y}^{II}$ are used for the inverse wavelet transform instead of d_{j,k_x,k_y}^{RR}, d_{j,k_x,k_y}^{RI}, d_{j,k_x,k_y}^{IR} and d_{j,k_x,k_y}^{II}.

However, it has been pointed out by many authors that the threshold λ obtained by Equation 5.79 is sometimes too large for image de-noising, although it can be applied to 2-D de-noising. One of the reasons for this is that the total number N of 2-D data tends to be large. Therefore, instead of Equation 5.79, we use λ expressed as follows:

$$
\lambda = K\sigma,
\tag{5.85}
$$

where, K decided by experimentation.

Figure 5.49a shows the image with Gaussian white noise added so that SMR = 6.0 dB, Figure 5.49b shows the de-noised image using the $m = 4$ spline wavelet (SP4), Figure 5.49c shows the de-noised image using the $m = 4, 3$ RI-spline wavelet (RI-SP) and Figure 5.49d shows de-noised image using smoothing filter (5×5 pixels). In the two cases of Figures 5.49b and 5.49c, de-noising was carried out by using Equation 5.78, and σ was decided using Equation 5.80. Figure 5.50 shows the root mean squared error (RMSE) between the de-noised and the original images plotted as a function of K. As shown in Figure 5.50, the lowest RMSE can be obtained around $K = 3$ for both the SP4 wavelet case and the RI-SP wavelet case. Thus we selected $K = 3$ for image de-noising. Comparing Figures 5.49b, 5.49c and 5.49d, it is clear that our method using the RI-SP wavelet has a better de-noising performance than that of the SP4 wavelet and the smoothing filter.

Here, we show some experimental results which were obtained with our method that uses the RI-SP wavelet applied to real medical images. Figure 5.51a shows an SMA thrombosis image. As this example shows, usual medical images need some sharpening. For sharpening images, amplifying

(a)   (b)

(c)   (d)

Fig. 5.49. Examples of de-noising results: (a) image with Gaussian noise, (b) de-noising result obtained by $m = 4$ spline wavelet, (c) de-noising result obtained by $m = 4, 3$ RI-spline wavelet, and (d) de-noising result obtained by smoothing filter (5×5 pixels)

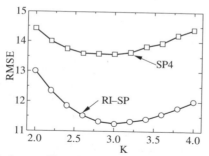

Fig. 5.50. Relation between K and RMSE obtained by the $m = 4$ spline wavelet and the $m = 4, 3$ RI-spline wavelet

wavelet coefficients of level -1 and level -2, which is equivalent to amplifying high frequency components, before reconstruction by the inverse wavelet transform is commonly done. However, at the same time, this sharpening method also amplifies noise, because a lot of noise is contained in high frequency components. We apply our de-noising method to the noise-amplified sharpened images. Figure 5.51b shows the de-noised image by using ordinary wavelet shrinkage applied to the sharpened image of Figure 5.51a. For sharpening, we used the 2-D DWT using a real mother wavelet, the SP4 in this case, then magnified the wavelet coefficients of level -1 and level -2 by four

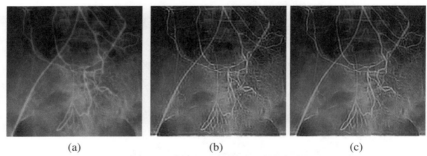

(a)   (b)   (c)

Fig. 5.51. De-noising result of SMA thrombosis using the $m = 4$ spline wavelet and the RI-spline wavelet: (a) original SMA thrombosis image, (b) sharpened and de-noised result obtained by using the $m = 4$ spline wavelet and (d) sharpened and de-noised result obtained by using the $m = 4, 3$ RI-spline wavelet

times. Figure 5.51c shows the de-noised image by our method, which uses 2-D CDWT with the RI-SP wavelet, applied to the sharpened image obtained the same way as Figure 5.51b. In these two cases, K in Equation 5.85 was selected as $K = 3$ according to Figure 5.51, and σ was decided using Equation 5.80. Comparing Figure 5.51c with Figure 5.51b, we see that in the de-noised image obtained by our method less distortions near edges are observed, which enables clearer images to be obtained by our method.

B. Removing Textile Texture

As is shown above, wavelet shrinkage is a simple but effective image de-noising method. However textures contain not only random components but also deterministic components [31]. So the method of setting up threshold λ differs greatly from the case of A [32]. However, after threshold λ has been set up, the processing removes a textile texture is the same as the case of A.

In order to determine the threshold λ, the "good sample" that does not include a defect is used first. The 2-D CDWT here is applied to the "good sample" that does not include a defect and the image, for example, Figure 5.25b that consisting of the TI coefficients expressed with Equation 5.72 is obtained. The TI coefficients then put in order according to the size of the value for each sub-band except the LL sub-band and the threshold λ is selected as the TI coefficients' value at 90% rank order from the largest value. Using the threshold λ obtained above, the textile texture of the textile surface image serving as a subject of examination is then removed by Equation 5.78. This should just from the method stated in A for each level. Hereafter, we call the image removed be done by applying the textile texture from the original image reconstructed image.

Once the textile textures are removed from the textile surfaces, the remaining inspection processes becomes a tractable problem. We use a simple statistical method as follows. First, we estimate σ_r in advance, which is the

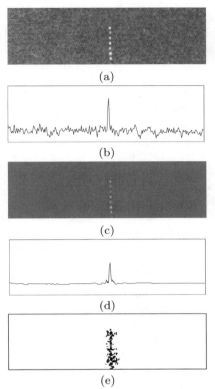

(a)

(b)

(c)

(d)

(e)

Fig. 5.52. Experimental results: (a) observed image, (b) profile of (a), (c) reconstructed image of (a), (d) profile of (c), and (e) detected defects of (a)

standard deviation of the distribution of the TI coefficient values in the reconstructed "good sample" image, which contains no defects. Here, the reconstructed image means the one from which the texture is removed. Using this σ_r, we apply thresholding with Equation 5.86. If the TI coefficient values b of the reconstructed images to be inspected lie in the range expressed in Equation 5.86 then they are marked white, otherwise they are marked black

$$\overline{u_b} - a\sqrt{2\log(N)}\,\sigma_r \leq b \leq \overline{u_b} + a\sqrt{2\log(N)}\,\sigma_r. \qquad (5.86)$$

In Equation 5.86, $\overline{u_b}$ is the mean of the TI coefficient values of the reconstructed images to be inspected. N means the total number of pixels, and a is an adjustable parameter. If the histogram distribution of the reconstructed "good sample" image can be approximated by a Gaussian distribution and the value of the parameter a is 1.0, the expected total number of pixels whose TI coefficient values do not lie in the range expressed in Equation 5.86 becomes less than 1 (pixel). Thus we can treat the pixels whose TI coefficient values do not lie in the range expressed in Equation 5.86 as outliers. If many pixels are classified as outliers, we can conclude that some defects exist on the textile surface. However, this rule holds in the ideal case. In actual environments, we sometimes need to adjust the parameter a according to experimental conditions.

In the experiment, the monochrome image of the size of 512×512 taken from the place distant from the front of a lens of 100 cm is used. This image corresponds to the 14.5cm domain on the actual textile surface. For lighting, two lights of high frequency lighting fluorescent light were used. Furthermore, we used a 6 level decomposition and the threshold λ_ss were selected as the TI coefficient value at the 95% rank order from the largest value. The adjustable parameter a was fixed at 1.6.

Figure 5.52a shows only a 256×76 portion of a textile surface image including a thread difference defect and the image size corresponds to the actual size. Figure 5.52c is a reconstruction image corresponding to Figure 5.52a. The defective partial detection result obtained by using Equation 5.86 in the image Figure 5.52c is shown in Figure 5.52e. It is clear that the defective portion is detected well. Figure 5.52b shows the brightness change of the horizontal axis section containing the defective portion of Figure 5.52a. Similarly, Figure 5.52d shows the brightness change of the same horizontal axis section of Figure. 5.52c. Comparing Figures 5.52b and 5.52d it often turns out that the texture information is removed, without spoiling the defective portion information. The example of this section shows that the wavelet degeneration extended by 2-D CDWT is effective not only in removal of the signal, which consists of a random component but also in removal of the signal containing both the deterministic component and the random component.

C. Fingerprint Analysis by 2-D CDWT

The effectiveness of the direction selectivity of the two dimensional CDWT was tested by analyzing a fingerprint. Figure 5.53 shows the fingerprint analysis results obtained by CDWT, where Figure 5.53a shows images of the fingerprints, and Figure 5.53b six directional components corresponding to each fingerprint. In Figure 5.53a, the sample A is clear, the sample A′ is the same as sample A although it is dirty, and sample B is different from sample A and sample A′. By comparing Figure 5.53b, we can observe that the pattern of each directional component of A′ are similar to that of A although A′ is dirty, and sample B is different from sample A and sample A′.

Furthermore, in the Figure 5.53b, only the coefficients of direction 75° are retained, and other coefficients are rewritten to be 0. These coefficients are used for reconstruction by the inverse transform. Corresponding to it, for the DWT by using RI-spline wavelet, only the coefficients of direction 90° are retained and other coefficients are rewritten to be 0. These coefficients are used for reconstruction by the inverse transform. Figure 5.54 shows the results obtained by CDWT and DWT, where Figure 5.54a shows components in the 75° direction that were extracted by the CDWT and Figure 5.54b

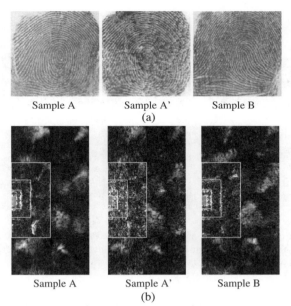

Sample A   Sample A'   Sample B
(a)

Sample A   Sample A'   Sample B
(b)

Fig. 5.53. Example of fingerprint direction analysis by 2-D CDWT: (a) samples of fingerprints (128×128), (b) analysis results obtained by CDWT on a scale of $1/2$

components in the 90° direction that were extracted by DWT. It is clear by comparing Figures 5.54a and 5.54b that the 2-D CDWT has a better capability of identifying the features of each fingerprint, also almost without influencing the bad picture in the state where a part of the fingerprint was blurred or rubbed.

5.5 Chapter Summary

Wavelet transform is a time-frequency method and has some desirable properties for nonstationary signal analysis and has received much attention. The wavelet transform uses the dilation b and translation a of a single wavelet function $\psi(t)$ called the *mother wavelet* (MW) to analyze all different finite energy signals. It can be divided into the continuous wavelet transform (CWT) and the discrete wavelet transform (DWT) based on the variables a and b, which are continuous values or discrete numbers. Many famous reference books on the subject have been published [4, 5].

However, when CWT and DWT are used in the manufacturing systems as a signal analysis method, there are still some problems. In the case of CWT, the following problems can be arise. 1) CWT is a convolution integral in the time domain, so the amount of computation is enormous and it is impossible to analyze the signals in real time. Moreover, as yet there is still no common

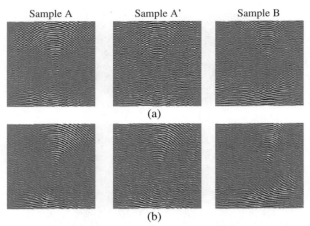

Fig. 5.54. Example of the fingerprint direction extracted by 2-D CDWT and 2-D DWT on level -1: (a) result obtained by DWT on level -1: LH (90°) and (b) obtained by CDWT on level -1: 75°

fast algorithm for CWT computation although it is an important technology for manufacturing systems. 2) CWT can show unsteady signal features clearly in the time-frequency plane, but it cannot quantitatively detect and evaluate its features at the same time because the common MW performs bandpass filtering. At same time, in the case of DWT, following problems can arise. 1) The transformed result obtained by DWT is not translation invariant. This means that shifts of the input signal generate undesirable changes in the wavelet coefficients. Thus DWT cannot catch features of the signals exactly. 2) DWT has poor direction selection in the Image. That is, the DWT can only obtain the mixed information of $+45^o$ and -45°, although each direction information is important for surface inspection.

Therefore, in this chapter, we focused on the problems shown above and discussed the following methods for improvement:

1. A fast algorithm in the frequency domain for improving the CWT's computation speed.
2. The wavelet instantaneous correlation (WIC) method by using the real signal mother wavelet (RMW), constructed from a real signal for detecting and evaluating abnormal signals quantitatively.
3. The complex discrete wavelet transform by using the real-imaginary spline wavelet (RI-spline wavelet) for improving the DWT's drawbacks such as the lack of translation invariance and poor direction selection.

Furthermore, we applied these methods to de-noising, abnormal detection, image processing and so on, and showed their effectiveness. The results in this chapter may contribute to improving the capability of the wavelet transform for manufacturing systems. Moreover, they are indicative of the future possibility of the wavelet transform as a useful signal and image processing tool.

References

1. Cohen L (1995) Time-frequency analysis. Prentice-Hill PTR, New Jersey
2. Chui C K (1992) An introduction to wavelets. Academic Press, New York
3. Coifman RR, Meyer Y and Wickerhauser (1992) Wavelet analysis and signal processing. In Ruski MB et al. (ed.) Wavelet and their applications, pp.153–178, Jones and Bartlett, Boston
4. Daubechies I (1992) Ten lectures on wavelets. SIAM, Philadelphia
5. Mallat SG (1999) A wavelet tour of signal processing. Academic Press, New York
6. Mallat SG (1989) A theory for multiresolution signal decomposition: the wavelet representation. IEEE Transaction on Pattern Analysis and Machine Intelligence, 11:674–693
7. Magarey JFA and Kingsbury NG (1998) Motion estimation using a complex-valued wavelet transform. IEEE Transaction on Signal Processing, 46:1069–1084
8. Kingsbury N (2001) Cpmplex wavelets for shift invariant analysis and filtering of signals. Journal of Applied and Computational Harmonic Analysis, 10:234–253
9. Zhang Z, Kawabata H, and Liu ZQ (2001) Electroencephalogram analysis using fast wavelet transform. International Journal of Computers in Biology and Medicine, 31:429–440
10. Zhang Z, Horihata S, Miyake T and Tomita E (2005) Knocking detection by complex wavelet instantaneous correlation. Proc. of the 13th International Pacific Conference on Automotive Engineering, pp.138–143
11. Zhang Z, Toda H, Fujiwara H and Ren F (2006) Translation invariant ri-spline wavelet and its application on de-noising. International Journal of Information Technology & Decision Making, 5:353–378
12. Holschneider M (1995) Wavelets, an Analysis tool. Oxford University Press
13. Zhang Z, kawabata H and Liu ZQ (2001) Nonstationary signal analysis using the RI-spline wavelet. Integrated Computer-Aided Engineering, 8:351–362
14. Unser M (1996) A practical guide to the implementation of the wavelet transform. In: Aldroubi A and Unser M (ed.) Wavelets in medicine and biology, pp.37–73, CRC Press
15. Shensa MJ (1992) The discrete wavelet transform: wedding the á trous and Mallat algorithms. IEEE Transactions on Signal processing, 40:2464–2482
16. Yamada M, and Ohkitani K (1991) An identification of energy casade in turbulence by orthonormal wavelet analysis. Progress Theoretical Physics. 86: 99–815
17. Maeda M, Yasui N, Kitagawa H and Horihata S (1996) An algorithm on fast wavelet transform/inverse transform and data compression for inverse wavelet transform. Proc. JSME 73th General Meeting, pp.141–142 (in Japanese)
18. Rioul O and Duhamel P (1992) Fast algorithms for discrete and continuous wavelet transform. IEEE Transactions on Information Theory, 38:569–586
19. Selesnick, IW (2001) Hilbert transform pairs of wavelet bases. IEEE Transactions on Signal Processing Letters, 8:170–173
20. Fernandes Felix CA, Selesnick, IW, Spaendonck Rutger LC van and Burrus CS (2003) Complex wavelet transforms with allpass filters. Signal Processing, 88:1689–1706

21. Peitge HO and Saupe D (1988) The science of fractal image. Springer, New York
22. Higuchi T (1989) Fractal analysis of time series. Proc. of Institute of Statistical Mathematics, 37:210–233 (in Japanese)
23. Heywood JB (1988) Internal combustion engine fundamentals. Mc-Graw Hill, New York
24. Samimy B, Rizzoni G and Leisenring K, (1995) Improved knocking detection by advanced signal processing. Special Publication SP-1086, Engine Management and Driveline Controls, pp.178–181 (SAE Paper No. 950845)
25. Donoho DL and Johnstone IM (1994) Ideal spatial adaptation by wavelet shrinkage. Biometrika, 81:425–455
26. Coifman RR and Donoho DL (1995) Translation invariant de-noising in wavelets and statistics. Lecture Notes in Statistics, pp.125-150, Springer Berlin
27. Romberg JK, Choi H and Baraniuk RG (1999) Translation invariant denoising using wavelet domain hidden Markov tree. In: Conference record of the 33rd asilomar conference on Signals, Systems and Computers, Pacific Grove, CA
28. Cohen I, Raz S and Malah D (1999) Translation invariant denoising using the minimum description length criterion. Signal Processing, 75:201–223
29. Mochimaru F, Fujimoto Y andIshikawa Y (2002) Detecting the fetal electro-cardiofram by wavelet theory-based methods. Progress in Biomedical Research, 7:185–193
30. Ercelebi E (2004) Electrocardiofram signals de-noising using lifting discrete wavelet transform. Compters in Biology and medicine, 34:479–493
31. Francos JM, Meiri AZ and Porat B (1993) A unified texture model based on a 2D world-like decomposition. IEEE Transactions on Signal Processing, 41:2665–2678
32. Fujiwara H, Zhang Z and Hashimoto K (2001) Toward automated inspection of textile surfaces: removing the textural information by using wavelet shrinkage. IEEE International Conference on Robotics and Automation (ICRA2001), pp.3529–3534
33. Meyer Y (1993) Wavelets, algorithms and applications. SIAM, Philadelphia

6

Integration of Information Systems

6.1 Introduction

Information systems have been playing an active role in manufacturing since the early days of inventory control systems and have grown quickly in the past 20 years. Nonetheless, the origin of modern manufacturing information systems goes back to the 1950s. At the beginning, the main purpose of these systems was to support financial work on the one hand and process control on the other. The functions implemented in financial systems included inventory control, purchase. Process control systems were analog controllers that implemented basic control logic to operate actuators such as valves and electric devices.

The financial information systems evolved into material requirements planning (MRP) , manufacture resource planning (MRPII) , and subsequently into the enterprise resource planning (ERP) systems. Nowadays, ERP systems fall into the broader category of Enterprise Systems which are designed to manage inventory levels and resources, plan production runs, drive execution and calculate costs of making products.

On the other side of the spectrum, process control systems evolved into programmable logic controllers (PLCs), distributed control systems (DCS), and modern supervision and control systems that replaced the old relay logic operator control panels. ¿From the start of the 1990s the necessity of connecting information systems, such as ERP, and equipment control systems became clear. For instance, real-time data from the factory floor to business decision makers has a significant impact on improving the efficiency of the supply chain and decreasing cycle times. Conversely, the availability of information such as capacities, material availability and labor assignments deeply influences the efficiency of tasks such as job sequencing, and equipment scheduling.

Then manufacturing execution system (MES) came into existence as an approach to link business systems and control systems. MES are now being used in discrete, batch and continuous manufacturing industries, including aerospace, automotive, semiconductor, pharmaceutical, chemical, and petro-chemical industries. MES systems or manufacturing systems in general are designed to carry out activities such as controlling the execution of produc-tion orders, ensuring that recipes and other procedures are followed, capturing consumption of raw materials and quality data.

The next challenge was to integrate the numerous functions between man-ufacturing systems and business systems. As can be seen from Figure 6.1, man-ufacturing and enterprise systems operate with very different time frames, and the data managed differs in the level of detail. On the one hand, enterprise-level applications such as global planning have to deal with data in the order of weeks to months. On the other hand, manufacturing systems obtain data from individual machines and equipment with time scales ranging from hours to seconds.

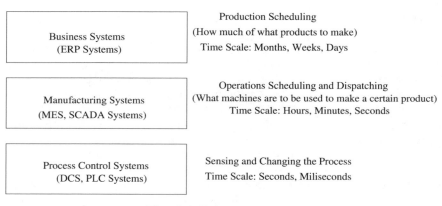

Fig. 6.1. Information systems

The increase of business complexity has added more requirements to infor-mation systems. It is not uncommon for an enterprise to deal with manufactur-ing operations on two or more separate sites managed by different companies. The enterprise may have its own business system that needs to be integrated with manufacturing systems from different vendors as in the example shown in Figure 6.2.

Enterprise and manufacturing systems are composed of modules that carry out individual functions (Figure 6.3). Consequently, developing interfaces be-tween the modules is complicated by the multiplicity of views of information. Enterprise applications such as ERP systems are concerned with information such as quantities and categories of resources (raw materials, equipment and personnel), and the amount of product produced. Scheduling systems require more specific information such as machine usage and batch recipes. Control systems require equipment connectivity information as well as individual mea-surements (such as temperature and pressure).

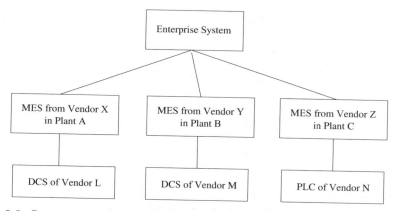

Fig. 6.2. Scenario of the integration between business and manufacturing systems

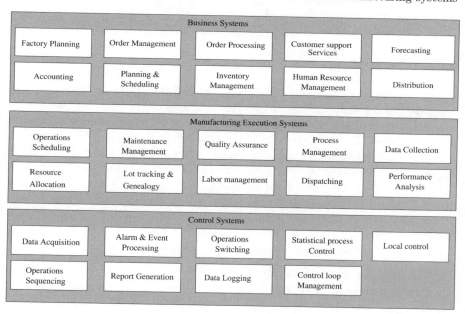

Fig. 6.3. Enterprise and manufacturing systems

Unfortunately, when the databases of these applications are developed from scratch they tend to be prescriptive (what will be) because the information models of the databases are developed so as to meet integration requirements imposed by either existing software or by functions to be carried out by software components. The more applications are included in the integration architecture, the more difficult is the integration. This situation may explain why average enterprise worldwide spends up to 40% of its IT budget on data integration [1].

6.2 Enterprise Systems

Enterprise systems are computer-based applications designed to process an organization's transactions and facilitate integrated and real-time planning, production, and customer response [2]. The following are some of the functions addressed by enterprise systems:

- What products should be made?
- How much of each product should be produced?
- What is the cost of producing each product?
- What are the resources to be allocated for producing each product?

Enterprise systems are complex applications that are usually built around a database that encompasses all the business data. For example, ERP software packages integrate in-house databases and legacy systems into an assembly with a global set of database tables .

6.3 MES Systems

The Manufacturing Enterprise Solutions Association (MESA International) proposes the following definition of MES:

"A manufacturing execution system (MES) is a dynamic information system that drives effective execution of manufacturing operations. Using current and accurate data, MES guides, triggers and reports on plant activities as events occur. The MES set of functions manages production operations from point of order release into manufacturing to point of product delivery into finished goods. MES provides mission critical information about production activities to others across the organization and supply chain via bi-directional communication."

The functions of MES systems are listed below:

1. Resource allocation and status.
2. Dispatching production orders.
3. Data collection/acquisition.
4. Quality management.
5. Maintenance management.
6. Performance analysis.
7. Operations scheduling.
8. Document control.
9. Labor management.
10. Process management.
11. Work-in progress (WIP) and lot tracking.

Manufacturing systems are not out-of-the-box software applications, rather they are composed of customizable modules, some of which are sold by vendors specializing in a certain areas such as quality assurance or maintenance management (Figure 6.3).

MES systems are becoming ubiquitous in production sites but the extent to which their integration capabilities contribute to their success is yet to be determined. The fact that major MES and ERP vendors provide holistic software solutions is a significant factor contributing to their success. These software solutions are defined in a top-down approach in which existing applications are replaced with modules in the ERP or MES system.

6.4 Integration Layers

Integration can be achieved by means of the use of one or more integration layers. The process integration layer defines flows of information between applications. The data integration deals with common terminology and data-structures shared between two or more applications. The main role of the lowest integration layer is to enable applications to call methods or services in an other application.

6.5 Integration Technologies

This section looks at integration technologies, each of which covers one or more integration layers.

6.5.1 Database Integration

Applications can exchange data by means of writing to and reading from the same database. In other words, a database is shared between two or more applications. This is possible by means of a lock that prevents others from modifying the same data at the same time. In other words, when an application locks a database record for write access, no other application can access that record for write until the lock is released.

Databases are typically developed in three stages:

- domain analysis
- information modeling
- physical design

Domain analysis defines what information is produced or consumed and by whom. IDEF0 is a systematic method to perform domain analysis developed by the United States Air Force as a result of the Air Force's Integrated Computer Aided Manufacturing (ICAM) program. Activity models can show which software tools or persons participate in the same activity. Activity modeling shows the information that is used or produced by an activity. Consequently, data requirements can be identified for producing an information model.

A rectangular box graphically represents each activity with arrows reading clockwise around the four sides as shown in Figure A.1 of Appendix A. These arrows are also referred to as ICOM (inputs, constraints or controls, outputs and mechanisms). Inputs represent the information used to produce the output of an activity. Constraints define the information that constrains and regulates an activity. Mechanisms represent the resources such as people or software tools that perform an activity.

Information modeling focuses on the development of information models that define the structure of the information that is to be shared between applications. Information models are composed of entities, attributes, and relationships among the entities.

The physical design of the database is done in terms of database tables along with their relationships, formats, and rules that constrain the input data. This activity is typically carried out using the software of the actual database.

6.5.2 Remote Procedure Calls

Remote procedure calls (RPC) is a technique that allows one application to request a service from an application located in another computer in a network without having to understand network details.

A. CORBA

CORBA (common object request broker architecture) is a kind of RPC architecture and infrastructure with which applications can interoperate locally or through the network (Figure 6.4). Applications play the roles of either servers or clients. A server has services that can be requested by a client through an IDL interface. Each server has one or more IDL interfaces defined in a language also called IDL. The IDL interface definition contains references to the actual services (methods and procedures) implemented in the server. To support these references, the specification of the IDL language includes mappings from IDL to many programming languages, including C, C++, Java, COBOL and Lisp. Using the standard protocol IIOP, a client can access remote services through the network.

B. COM/DCOM

DCOM is a kind of RPC architecture based on Microsoft RPC, which is compliant with DCE RPC (distributed computing environment RPC) defined by Open Software Foundation (OSF). Some features include the ability for an object to dynamically discover the interfaces implemented by other objects and a mechanism to uniquely identify objects and their interfaces.

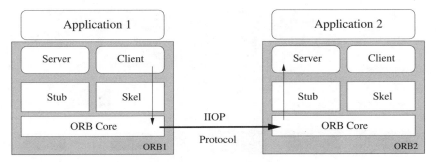

Fig. 6.4. CORBA architecture

C. JRMI

Java remote method invocation (JRMI) is an RPC architecture with which programs written in the Java language interact with programs running on other networked computers.

In a JRMI architecture, a client is supplied with the interface of methods available from the remote server. Each server has one or more JRMI interfaces that are used to inform clients what services are available and what data is to be returned from the services. The JRMI interface definition contains references to the actual services (methods and procedures) implemented in the server.

6.5.3 OPC

OPC is a standard set of interfaces, properties, and methods for use in process-control and manufacturing applications based on either DCOM or Web services (Figure 6.5).

Specifically, OPC provides a common interface for communicating with distributed control systems (DCSs), process supervisory control systems, PLCs, and other plant devices.

6.5.4 Publish and Subscribe

Publish and subscribe architectures are characterized by asynchronous integration between applications that are loosely-coupled. The infrastructure that facilitates the integration is known as message oriented middleware (MOM) , or message queuing.

Applications are classified as publishers or subscribers. Publishers post messages without explicitly specifying recipients or having knowledge of intended recipients. Subscribers are applications that receive messages of the kind that the subscriber has registered. Messages are typically encoded in a predefined format including XML [3] . XML is a format that resembles HTML

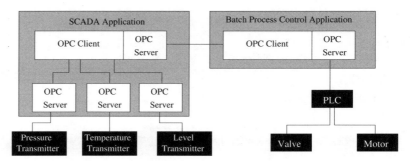

Fig. 6.5. Example of an OPC architecture

(the format used to build Web pages). XML can be identified by components of the format called tags that are delimited by the symbols < and > as shown in the bill of materials example of Sect. 6.5.5.

The message oriented middleware that mediates between publishers and subscribers manages a distribution mechanism that is organized in topics. In other words, the distribution mechanism delivers messages to the appropriate subscriber. A subscriber subscribes to a queue by expressing interest in messages that match a given topic. Publishers can post messages to one or more queues.

6.5.5 Web Services

A Web service is a software system designed to support interoperable machine-to-machine interaction over a network. A Web service has an interface defined in computer-processable format such as WSDL. Applications can request specific actions from the Web service using SOAP messages.

SOAP (simple object access protocol) is a protocol for exchanging messages between applications similar to publish-and-subscribe message formats. SOAP messages are encoded in XML.

Below is an example of a SOAP message sent by an application requesting the bill of materials (BOM) from a fictitious manufacturing Web service. The bill of materials corresponds to a bicycle with product identification number b789.

```
<soap:Envelope
  xmlns:soap="http://schemas.xmlsoap.org/soap/envelope/">
  <soap:Body>
    <getBillOfMaterials
      xmlns="http://manufacturing.example.com/bom">
      <productID>b789</productID>
    </getBillOfMaterials>
  </soap:Body>
</soap:Envelope>
```

The response of the Web Service is shown below.

```
<soap:Envelope
  xmlns:soap="http://schemas.xmlsoap.org/soap/envelope/">
  <soap:Body>
    <getBillOfMaterialsResponse
      xmlns="http://warehouse.example.com/ws">
      <getBillOfMaterialsResult>
        <productID>b789</productID>
        <partName>wheels</partName>
        <partName>frame</partName>
        <partName>handlebars</partName>
        <partName>seat</partName>
        <partName>pedals</partName>
        <partName>lights</partName>
        <partName>trim</partName>
        ...
      </getBillOfMaterialsResult>
    </getBillOfMaterialsResponse>
  </soap:Body>
</soap:Envelope>
```

6.6 Multi-agent Systems

The concept of "agency" is derived from research in the area of artificial intelligence and has evolved during the last 30 years. An agent as defined by Shoham is "a software entity which functions continuously and autonomously in a particular environment, often inhabited by other agents and processes" [4].

Agents are said to be autonomous in the sense that they can operate without the direct intervention of the user or other agents. This is the main distinction from the previous integration approaches. A server that offers a function to be called from an external application is allowing the client to exert direct control over one of its internal behaviors. However, agent interactions are mostly peer-to-peer, so the agent version of a server will accept "requests" to perform an action and it will be up to the agent to decide whether the action is executed, as well as the order of the execution of the action with respect to other actions in the agenda kept by the agent. A group of agents can, in fact, act cooperatively in order to carry out the activities that satisfy global requirements such as makespan constraints in scheduling.

The goal-directed behavior that agents exhibit is the result of having their own goals in order to act. In other words, agents not only respond to the environment, they have the ability to work towards their goals by taking the initiative. If part of the system is required to interact with an environment that is not entirely predictable, a static list of goals is not enough. For example, if an unexpected fault occurs during the analysis of the startup of the plant, the original plan for starting up the plant becomes an invalid result. As a responsive entity an agent has the property of responding to the environment with an emergent behavior (including reacting from unforeseen incidents).

Of particular importance during on-line operations is that agents should respond in a real-time fashion, in which changes in the environment do not always present a predictable behavior.

Agents communicate by exchanging messages that follow a standard structure. KQML was the first standard agent communication language and was developed by the ARPA supported Knowledge Sharing Effort consortium [5]. A KQML message consists of a performative and a number of parameters. Below is how an agent would encode the BOM request described in the Web service example.

```
(insert
 :contents
    <getBillOfMaterials
      xmlns="http://manufacturing.example.com/bom">
      <productID>b789</productID>
    </getBillOfMaterials>
 :language bomxml
 :ontology manufacturing_ontology
 :receiver manufacturing_agent
 :reply-with nil
 :sender assembly_agent
 :kqml-msg-id 5579+perseus+1201)
```

The performatives describe the intention and attitudes of an agent towards the information that is to be communicated, some of which are listed in Table 6.1.

6.6.1 FIPA: A Standard for Agent Systems

FIPA (Foundation for Intelligent Physical Agents) is an organization that has developed a collection of standards for agent management, agent communication and software infrastructure. Although originally formed as an European initiative in 1996, FIPA has also become an IEEE Computer Society standard [6].

A. Agent Architecture

The FIPA agent architecture has the following mandatory components:

- agent platform (AP)
- agent

Table 6.1. Basic KQML performatives

Category	Performatives
Basic informational performatives	tell, untell, deny
Basic query performatives	evaluate, reply, ask-if, ask-one, ask-all, sorry
Factual performatives	insert, uninsert, delete-one, delete-all, undelete
Multi-response query performatives	stream-about, stream-all, generator
Basic effector performatives	achieve, unachieve
Intervention performatives	next, ready, eos, standby, rest, discard
Capability definition	performatives advertise
Notification performatives	subscribe, monitor
Networking performatives	register, unregister, forward, broadcast, pipe, break
Facilitation performatives	broker-one, broker-all, recommend-one, recommend-all, recruit-one, recruit-all

- directory facilitator (DF)
- agent management system (AMS)
- message transport system

The agent platform is the combination of hardware and software where agents can be deployed. An AP consists of one or more agents, one or more directory facilitators, an agent management system and a message transport system. Each AP runs on one or more machines, each with its own operating system and all running a FIPA-compliant agent support software.

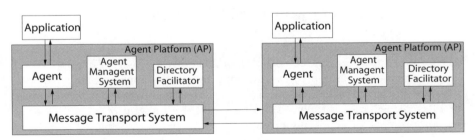

Fig. 6.6. FIPA agent management model

The agent in FIPA is defined as an autonomous entity that performs one or more services by using communication capabilities.

The directory facilitator (DF) is a type of agent that provides yellow page services similar to those of UDDI. In order to advertise their services, agents register with the DF by providing their name, location, and service description. With service information stored in the DF, agents can query the DF to find agents that match a certain service.

The agent management system (AMS) implements functions such as the creation and termination of agents, registration of agents on an AP and the management of the migration of agents from an AP to another AP. The most basic task of the AMS is to provide an agent name service (ANS), which is a kind of white pages containing network-related information about the agents registered on an AP. Each entry in the ANS includes the unique name of the agent and its network address for the AP. Each agent has an identifier composed of a unique name and the addresses of the platform where the agent resides.

The message transport system is responsible for routing messages between agents on an AP and between APs. The default communication protocol is the Internet inter-orb protocol (IIOP). However, other communication protocols such as HTTP are also permitted.

B. Agent Communication Language (ACL)

Agents exchange messages using an agent communication language (ACL). The structure of the message is similar to that of KQML. The ACL defines the structure of a message using a series of parameters and their values.

KQML performatives and FIPA ACL communicative acts are based on ideas from speech act theory. Speech act theories attempt to describe how people communicate their goals and intentions. Like the KQML performatives, the FIPA communicative acts describe the intention and attitudes of an agent in regards to the content of the message that is exchanged. Table 6.3 shows some of the standard FIPA communicative acts.

6.7 Applications of Multi-agent Systems in Manufacturing

According to reviews conducted by Shen and Norrie [7] and Tang and Wong [8], a number of research projects involving agents in manufacturing have been reported in the literature. Applications include scheduling, control, assembly line design, robotics, supply chain and enterprise integration. The following section presents some specific examples.

6.7.1 Multi-agent System Example

A matchmaking architecture is a computer environment made up of agents that communicate through Internet so that process designers and policy makers can search knowledge sources distributed geographically.

An example of such architecture is shown in Figure 6.7. The objective of this multi-agent architecture is to provide the means to find industrial processes that convert raw materials into desired products. A process agent (PA) manages key aspects about each individual process including the type of

Table 6.2. Parameters of an ACL message

Parameter	Description
performative	The type of the communicative act of the ACL message.
sender	The identity of the sender of the message, that is, the name of the agent of the communicative act.
receiver	The identity of the intended recipients of the message.
reply-to	This parameter indicates that subsequent messages in this conversation thread are to be directed to the agent named in the reply-to parameter, instead of to the agent named in the sender parameter.
content	The content of the message; equivalently denotes the object of the action. The meaning of the content of any ACL message is intended to be interpreted by the receiver of the message. This is particularly relevant for instance when referring to referential expressions, whose interpretation might be different for the sender and the receiver.
language	The language in which the content parameter is expressed.
encoding	The specific encoding of the content language expression.
ontology	The ontology(s) used to give a meaning to the symbols in the content expression .
protocol	The interaction protocol that the sending agent is employing with this ACL message.
conversation-id	Introduces an expression (a conversation identifier) which is used to identify the ongoing sequence of communicative acts that together form a conversation.
replywith	The reply-with parameter identifies a message that follows a conversation thread in a situation where multiple dialogs occur simultaneously
inreplyto	An expression that references an earlier action to which the message is a reply.
replyby	A time and/or date expression which indicates the latest time by which the sending agent would like to receive a reply.

product and feedstock constraints. PAs advertise their services with the directory facilitator (DF) who manages the yellow pages for all the environment's agents. PAs accept messages from process requesters (PRs) to evaluate the degree of matching between the process requirements and the capabilities of the process known by the PAs. A process requester can obtain information about PAs by contacting the DF. Decision makers interact with a PR using its graphical user interface. Process requirements are defined with the user interface by means of specifying the characteristics of the waste (*e.g.*, demolition wood) and desired products (*e.g.*, synthesis gas).

Table 6.3. FIPA communicative acts

Communicative act	Description
accept-proposal	The action of accepting a previously submitted proposal to perform an action.
agree	The action of agreeing to perform some action, possibly in the future.
cancel	The action of one agent informing another agent that the first agent no longer has the intention that the second agent performs some action.
cfp	Call for proposal. The action of calling for proposals to perform a given action.
confirm	The sender informs the receiver that a given proposition is true, where the receiver is known to be uncertain about the proposition.
disconfirm	The sender informs the receiver that a given proposition is false, where the receiver is known to believe, or believe it likely that, the proposition is true.
failure	The action of telling another agent that an action was attempted but the attempt failed.
inform	The sender informs the receiver that a given proposition is true.
not-understood	The sender of the act (for example, agent_a) informs the receiver (for example, agent_b) that it perceived that agent_b performed some action, but that agent_a did not understand what agent_b just did. For example when agent_a tells agent_b that agent_a did not understand the message that agent_b has just sent to agent_a.
propose	The action of submitting a proposal to perform a certain action, given certain preconditions.
query-if	The action of asking another agent whether or not a given proposition is true.
refuse	The action of refusing to perform a given action, and explaining the reason for the refusal.
reject	The action of rejecting a proposal to perform some action during a negotiation.
request	The sender requests the receiver to perform some action. One example of the use of the request act is to request the receiver to perform another communicative act.
subscribe	The act of requesting a persistent intention to notify the sender of the value of a reference, and to notify again whenever the object identified by the reference changes.

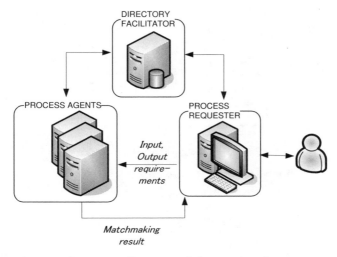

Fig. 6.7. System architecture of the matchmaking system

In order for agents to interoperate, ontologies are developed that define things such as substances, physical quantities, and units of measure. The capabilities of a given indutrial process are specified as a series of constraints on the allowed feedstock materials, and about the kind of products that can be obtained. Constraints are encoded in knowledge interchange format (KIF) which in reality represents queries to the agent knowledge base.

The prototype was programmed in Java using the JADE library for distributed agent applications and the JTP inference system [9].

JADE (Java agent development framework) is a software framework. JADE provides a Java library that can be used to implement multi-agent systems. JADE uses a middle-ware that complies with the FIPA specifications. The agent platform can be distributed across machines. It also provides tools for monitoring and configuration.

In the JADE runtime environment, agent communication follows the FIPA standard described in Sect. 6.6.1.

Messages are encoded in FIPA ACL . A message contains a number of parameters such as performative, sender, receiver, content, language and ontology . The performative defines the declarative act. The matchmaking environment implements the request, query-ref and inform performatives.

A typical exchange of messages is shown in Figure 6.8. PAs advertise their services with the DF by sending a fipa-request message with the registration request in the content of the message. Also, a PR can make use of the yellow page services of a DF by sending a fipa-request message. After getting the list of all available PAs, a PR prepares a list of feedstock requirements and product specifications and submit this information to PAs by means of a fipa-query-ref message. Each PA then sends a numeric score that represents the

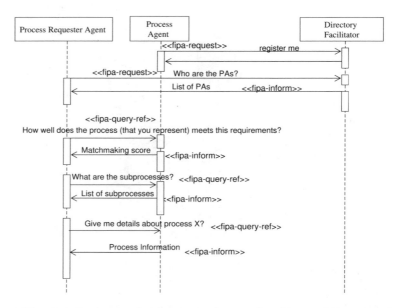

Fig. 6.8. Sequence of messages in the matchmaking architecture

degree of matching. Similar communication acts are used for obtaining the classes of sub-processes used in by the process in the profile of the PA.

6.8 Standard Reference Models

Standard reference models define domain-specific terminology and data structures that serve as an architecture for physical databases and as a basis for planning and implementing the integration.

6.8.1 ISO TC184

In the area of enterprise modeling and integration, the International Organization for Standardization Technical Committee 184 (ISO TC184) has been active in developing standards concerning discrete part manufacturing and encompassing the application of multiple industries including machines and equipment and chemicals. In regards to manufacturing integration, ISO TC184 activities are centered in two subcommittees: the TC184/SC 4 (industrial data) and TC 184/SC 5 (architecture, communications and integration frameworks). Some standards developed by TC 184/SC 4 are:

- ISO 10303 – Standard for the exchange of product model data, also known as STEP
- ISO 15531 – Manufacturing management data exchange (MANDATE)
- ISO 13584 – Parts library
- ISO 18629 – Process specification language (PSL)

ISO 10303, also known as the standard for the exchange of product model data (STEP), is a group of standards to represent and exchange product data on in a computer-processable format along the life cycle of a product. The range of products in STEP extends to printed circuit boards, cars, aircraft, ships, buildings, and process plants.

At the time of publishing this book, IEC/ISO 62264 is the only standard that defines the interface between production control and business systems.

6.9 IEC/ISO 62264

The IEC/ISO 6224 is a standard reference model for the integration between enterprise and control systems. The IEC/ISO 62264 is better known as the S95 standard, as it was originally developed by a group of system vendors and system integrators in the ISA (Instrumentation, Systems and Automation Society) SP95 committee. S95 is based on the Purdue reference model ([10]), the MESA international functional model and the equipment hierarchy model from the IEC 61512-1 standard ([11]).

The scope of the standard is described specifically by using a functional hierarchy model (Figure 6.9). Level 4 is concerned with basic production planning and scheduling functions as carried out by ERP, MRP or MRPII systems. Level 3 is concerned with functions implemented in MES systems. Levels 0, 1, and 2 refer to process control activities such as those carried out by PLCs and DCS systems. The IEC/ISO 6224 covers level 3 and some of level 4 activities.

Activities are carried out according to a specified part–whole relations for the manufacturing facility. These relations are defined in the equipment hierarchy (Figure 6.10).

There are three kinds of resources defined in the standard, namely personnel, material, and equipment.

Production activities are modeled by means of the production schedule, production performance, process segment, and production capacity.

Production schedule defines one or more production requests. It also defines the start and end times of the production and the location (enterprise, site, area, *etc.*) where the production is to take place.

Production performance is a collection of production responses. A production response is an item reported to the business system that contains information on the actual resources used until the end of the production.

Process segment defines a logical grouping of resources required to carry out a production step (an activity) at the level of detail required for planning

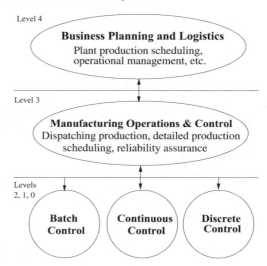

Fig. 6.9. IEC/ISO 62264 functional hierarchy

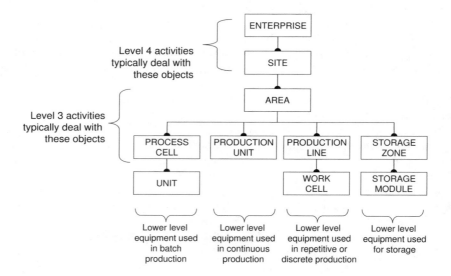

Fig. 6.10. IEC/ISO 62264 equipment hierarchy

or costing. Let us assume that a pharmaceutical factory produces pill packs and the accounting requires tracking three intermediate materials: active ingredient, pills and pill packs. Consequently, there are three process segments that are required by accounting: process_segment_1(make active ingredient), process_segment_2(make pills), and process_segment_3(package pills).

Production capacity is a collection of capabilities of resources (personnel, equipment, material) and process segments for a given period of time. Each resource is marked as committed, available or unattainable.

Information required to produce a given product is given by the product definition. The product definition for producing the bicycle in the SOAP example is shown below in B2MML, which is the XML encoding of the IEC/ISO 62264 models . Note that the manufacturing bill element (ManufacturingBill) is used to specify a material (part) needed to produce the product and its required quantity.

```
<ProductDefinition>
  <ID>b789</ID>
    <ManufacturingBill>
      <MaterialClassID>wheels</MaterialClassID>
      <Quantity>2</Quantity>
    </ManufacturingBill>
    <ManufacturingBill>
      <MaterialClassID>frame</MaterialClassID>
      <Quantity>1</Quantity>
    </ManufacturingBill>
    <ManufacturingBill>
      <MaterialClassID>handlebars</MaterialClassID>
      <Quantity>1</Quantity>
    </ManufacturingBill>
    <ManufacturingBill>
      <MaterialClassID>seat</MaterialClassID>
      <Quantity>1</Quantity>
    </ManufacturingBill>
    <ManufacturingBill>
      <MaterialClassID>pedals</MaterialClassID>
      <Quantity>2</Quantity>
    </ManufacturingBill>
    <ManufacturingBill>
      <MaterialClassID>lights</MaterialClassID>
      <Quantity>2</Quantity>
    </ManufacturingBill>
    ...
</ProductDefinition>
```

6.10 Formal Languages

A formal language is a set of lexical units and rules (syntax) required to represent application-independent data and knowledge. Formal languages are normally managed by standardization bodies so as to support information sharing and exchange in large user communities. In order to be useful in information systems integration, a formal language has to be both human and machine-processable. Information that is represented according to the syntactic rules of a formal language is typically encoded in a neutral format that is computer-processable (thus allowing this information to be exchanged among applications). XML (See Sect. 6.5.4) is an example of such a neutral format.

6.10.1 EXPRESS

Product data models in STEP are specified in EXPRESS (ISO 10303-11), a formal language that is based on entity–attribute-relationship languages and ideas from object oriented methods [12]. EXPRESS is defined in ISO 10303-11:1994. EXPRESS-G is a graphical language that provides a subset of the lexical modeling capabilities of EXPRESS as defined in Annex D of ISO 10303-11:1994. EXPRESS has also been adopted by many projects others than STEP. Among these one can find the Electronic Design Interchange Format standards and in the Petrotechnical Open Software Corporation's standards.

6.10.2 Ontology Languages

Whilst useful in many applications, information models in EXPRESS cannot be used directly in knowledge-based applications that require high expressive semantic content. On the other hand, a number of ontology languages have been developed with a variety of expressivity and robustness, including the formal languaged called OWL. The following example illustrates some capabilities of the use of models represented in Semantic Web languages. Let us assume we have an ontology for processes that defines a process as something that can be composed of other processes through the sub-process property. This can be represented in OWL as follows:

```
<owl:Class rdf:about="#PhysicalThing">
  <rdfs:subClassOf>
    <daml:Class rdf:about="#Physical"/>
  </rdfs:subClassOf>
</owl:Class>

<owl:ObjectProperty rdf:about="#part">
        <rdfs:domain>
                <daml:Class rdf:about="#PhysicalThing"/>
```

```
        </rdfs:domain>
        <rdfs:range>
            <daml:Class rdf:about="#PhysicalThing"/>
        </rdfs:range>
</owl:ObjectProperty>
```

```
<owl:TransitiveProperty rdf:about="#part"/>
```

6.10.3 OWL

OWL is an ontology language for the Web that provides modeling constructs to represent knowledge with a formal semantics [13]. OWL was developed by the World Wide Web Consortium (W3C) Web Ontology Working Group [14] and is being used to encode knowledge and enable interoperability in distributed computer systems [15]. The most fundamental concept in ontologies is that things can be grouped together as a set called class. The subClassOf relation is used to describe specializations of a more generic class. A class can be defined in terms of the properties that characterize it. For example, if we assert that every centrifugal pump is a device that contains an impeller, the definition of centrifugal pump can be represented in OWL as follows:

```
(Class centrifugal\_pump
    (subClassOf pump)
(subClassOf
(Restriction composition_of_individual
                (someValuesFrom impeller))))
```

which is equivalent to the following XML serialization

```
<owl:Class rdf:ID="centrifugal\_pump">
    <rdfs:subClassOf rdf:resource="#pump"/>
    <rdfs:subClassOf>
      <owl:Restriction>
        <owl:onProperty
          rdf:resource="iso15926&;composition_of_individual"/>
        <owl:someValuesFrom rdf:resource="#impeller"/>
      </owl:Restriction>
    </rdfs:subClassOf>
</owl:Class>
```

OWL provides constructs for defining relations in terms of their domains and ranges. The domain definition specifies the class to which the property belongs. Range definitions specify either OWL classes or externally-defined data types such as strings or integers. OWL uses the term Property to refer to relations.

Cardinality restrictions can be used to specify the exact number of values that should be on a specific relation of a given class. For example, a centrifugal pump can be defined as a pump that has at least one impeller.

A relation can be declared as transitive, symmetric, functional or inverse of another property. If a relation R is transitive, and R relates A to B, and B is related to C via R then A is related to C via R. For example, if the plate finned tube 123 is part of intercooler x and intercooler x is part of multi-stage compressor y then 123 is also part of y. A relation R is symmetric if when A is related to B then B is related to A in the same way. FunctionalProperty is a special type of relation such that for each thing in its domain, there is a single thing in its range. If some FunctionalProperty relates A to B then its inverse relation will link B to A. For example, if the relation contains is defined as FunctionalProperty then (contains tank-1 batch-1) is equivalent to (contained_in batch-1 tank-1) when contains is declared as an inverse relation of contained_in.

OWL provides constructs to define individuals (members of a class) such as those for describing which objects belong to which classes, the specific property values and whether two objects are the same or distinct. The prefixes owl, rdf, and rdfs are used to denote the namespaces where the OWL, RDF, and RDFS modeling constructs are respectively defined. Similar prefixes are also used to avoid name clashes, allowing multiple uses of a term in different contexts. For example, mil:tank and equip:tank can be used in an ontology to refer to a military tank and an equipment tank respectively.

OWL has the following benefits:

- Knowledge represented in OWL can be processed by a number of inference software packages.
- Support of the creation of reusable libraries.
- A variety of publicly available tools for editing and syntax checking.

6.10.4 Matchmaking Agents Revisited

In the matchmaking environment of Sect.6.7.1, queries to the ontology are passed to JTP (Java theorem prover), which is a reasoning system that can derive inferences from knowledge encoded in the OWL language. JTP is composed of a number of reasoners that implement algorithms such as generalized modus ponens, backward chaining, and forward chaining, and unification [16].

JTP translates each OWL statement into a KIF sentence of the form (PropertyValue Value Predicate Subject Object). Then it simplifies those KIF sentences using a series of axioms that define OWL semantics. OWL statements are finally converted to the form (Predicate Subject Object). Queries are formulated in KIF, where variables are preceded by a question mark.

All agents have copies of the upper and domain ontologies that can be retrieved from an Internet server. PRs use JTP in a number of ways. For example, the PRs can list the classes of biomass materials as in Figure 6.11, which are used by the decision maker to define a search session. In this example, the list of classes is obtained by querying the JTP knowledge base with the following query:

```
(rdfs:subClassOf ?x bio:compound)
```

This means that programming code of the agent remains unchanged even when new classes are added to the ontology file. JTP is also used to dynamically present user interfaces based on the information selected by the user. For example, if the decision maker is to define the water_content of a feedstock the PR presents a screen for entering the value and unit of measure associated to the mass_quantity. However, if the decision maker defines the phase of the feedstock then the PR presents a screen for specifying whether it is solid, liquid or gas. Again, there is no need to modify the agent's code if new units of measure or new properties are added to the ontology.

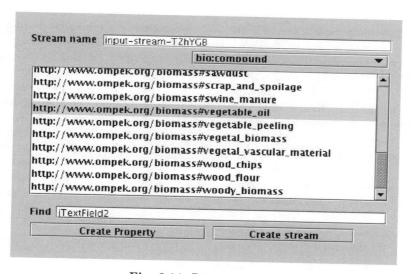

Fig. 6.11. Biomass classes

6.11 Upper Ontologies

Upper ontologies define domain-independent concepts such as physical objects, activities, mereological and topological relations from which more specific classes and relations can be defined. Examples of upper ontologies are SUMO [17], Sowa upper ontology [18], Dolce [19], and ISO 15926-2 [20] . Engineers can start by identifying key concepts by means of activity modeling, use cases and competency questions. This concepts are then defined based on the more general concepts provided by the upper ontology.

6.11.1 ISO 15926

ISO 15926-2-2003 is founded on an explicit metaphysical view of the world known as four–dimensionalism. In four–dimensionalism, objects are extended in space as well as in time, rather than being wholly present at each point in time, and passing through time. An implication of this is that the whole– part relation applies equally to time as it does with respect to space. For example, if a steel bar is made into a pipe then the pipe and the steel bar represent a single object. In other words, a spatio-temporal part of the steel bar coincides with the pipe and this implies that they are both the same object for that period of time. This is intuitive if we think that the subatomic particles of the pipe overlap the steel bar. Information systems have to support the evolution of data over time. For example, let us assume that a motor was specified and identified as M-100 so as to be installed as part of a conveyor. Some time later, the conveyor manufacturer delivers a conveyor that includes a motor with serial number 1234 that meets the design specifications of M-100. After a period of operation motor 1234 fails. Therefore, maintenance decides to replace it with motor 9876. This situation can be easily modeled using the concept of temporal parts as shown in Figure 6.12. ISO 15926 2:2003 defines the class functional physical_object to define things such as motor M-100 which have functional, rather than material continuity as their basis for identity. In order to say that motor 1234 is installed in a conveyor as M-100, M-100 is defined as consisting of S-1 (temporal part of 1234). In other words, S-1 is a temporal part of 1234 but is also a temporal part of M-100. In fact, because S-1 and P-101 have the same spatio-temporal extent they represent the same thing. Similarly, after a period of operation 1234 was removed and pump 9876 took its place. In this case, S-2 (temporal part of 9876) becomes a temporal part of P-101. Objects such as P-101 are known as replaceable parts, which is a concept common in artifacts in many engineering fields such as the process, automobile, and aerospace industries [21].

6.11.2 Connectivity and Composition

Part–whole relations of an object, which means that a component can be decomposed into parts or subcomponents that in turn can be decomposed into other components are defined by means of composition_of_individual and its subproperties. composition_of_individual is transitive. Subproperties of composition_of_individual include containment_of_individual (used to represent things that are inside others) and relative_location (used to locate objects on a particular place).

The following code shows a bicycle and its handlebars.

```
<owl:Class rdf:ID="bicycle">
  <rdfs:subClassOf rdf:resource="#physical_object"/>
```

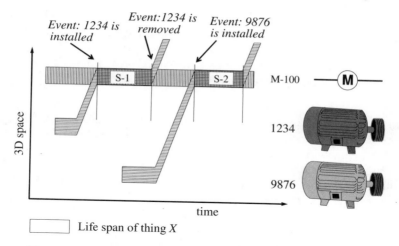

Fig. 6.12. Motor M-100 and its temporal parts 1234 and 9876

```
</owl:Class>

<owl:Class rdf:ID="handlebar">
  <rdfs:subClassOf rdf:resource="#physical_object"/>
</owl:Class>

...

<bicycle rdf:ID="b123">
  <composition_of_individual rdf:resource="handlebars1"/>
  ...
</bicycle>

<handlebars rdf:ID="handlebars1"/>

...
```

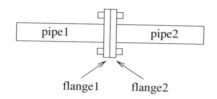

Fig. 6.13. Pipes connected by flanges

Connectivity between objects is based on connection_of_individual, which is defined as symmetric and transitive. For example, the symmetric character of the relation, allows us to infer that flange1 is connected to flange2, provided that pipe1 is connected to pipe2 in Figure 6.13. The definition of connection_of_individual and the topological description of the pipes are shown below:

```
<owl:ObjectProperty rdf:about="connection_of_individual">
    <rdfs:domain rdf:resource="#possible_individual"/>
    <rdfs:range rdf:resource="#possible_individual"/>
    <rdf:type rdf:resource=
    "http://www.w3.org/2002/07/owl#SymmetricProperty"/>
</owl:ObjectProperty>

<pipe rdf:ID="pipe1">
  <connection_of_individual rdf:resource="flange1"/>
</pipe>

<pipe rdf:ID="pipe2">
  <connection_of_individual rdf:resource="#flange2"/>
</pipe>

<flange rdf:ID="flange1">
  <connection_of_individual rdf:resource="flange2"/>
</flange>

<flange rdf:ID="flange2"/>
```

Using the axioms of transitiveness, an inference engine can conclude that pipe1 and pipe2 are connected because their flanges flange1 and flange2 are connected.

6.11.3 Physical Quantities

In September 1999, NASA was managing a mission to the planet Mars in order to study the Martian weather and climate by means of putting in orbit the Mars Climate Orbiter. Scientists at NASA's Jet Propulsion Laboratory in Pasadena, California received the thrust data from Lockheed Martin Astronautics in Denver (the spacecraft manufacturer). The data were expressed in Newtons, while the software control had an internal representation of this unit of measure in pounds of force. Units were not part of the input data and consequently the engineers assumed that the inputs were in Newtons. The loss of the Mars Climate Orbiter was caused by engineers who assumed the wrong units of measure. The error caused the spacecraft to enter the Martian atmosphere at about 57 km instead of the planned 140–150 km.

The corollary of this lesson is that a number of alone is not and will never be a physical quantity. Information systems must have a way to make this distinction. To understand the challenges involved in this quest let us try to model the force data of the Orbiter's thrust. The simplest way to solve this problem is to define objects with an attribute that represent a physical quantity as shown in Figure 6.14. The drawback of this approach is that the information system is not aware of the use of units of measure, as these are implicit in the name of the attribute.

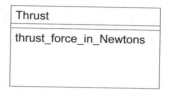

Fig. 6.14. Units of measure: approach 1

The second approach consists in defining two attributes, one representing the magnitude (the number) and another representing the unit of measure. This approach is more flexible, as a variety of units of measure for force might be chosen. However, again the information system is not aware of the relationship between the number in the magnitude attribute and the unit of measure (Figure 6.15).

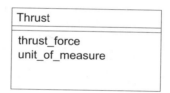

Fig. 6.15. Units of measure: approach 2

Another drawback common to both approaches is of ontological nature. Because attributes are what distinguishes instances of the same class, a thrust with 20 Newtons would be considered as a different instance from a thrust with 30 Newtons, while it was assumed that the same thrust can have different thrust forces along the life-cycle of the device.

Physical objects and processes should not use physical quantities (3 kg, 5 m, *etc.*) as attributes because a physical quantity is not an inherent property of an object [22]. For example, the setpoint of a temperature controller TIC_01 (a physical object) at 800 Kelvin should not be represented as an attribute (a relationship in ontology terms) because there is nothing intrinsic about 800

Kelvin that says it is the setpoint of TIC_01. "800 Kelvin" is just the extent of the temperature quantity to which the temperature set point refers.

The mapping between a controller and a temperature quantity can be defined as an instance of class_of_indirect_property. The class_of_indirect-_property is implemented as a subclass of owl:FunctionalProperty, whose domain is given by members of class_of_individual and whose range is given by members of property_space. temperature_setpoint is thus a relation whose range refers to instances of temperature_quantity. temperature_quantity is an instance of property_space, which makes it both a class and an instance. Furthermore, property_space is a subclass of class_of_property, which means that temperature_quantity is also an instance of class_of_property as shown in the code below. The OWL code also states that controller TIC_01 has temporal part whose setpoint is 800 Kelvin. The mapping between the value of 800 and the property is done by means of a property _quantification. A property_quantification is a functional mapping whose members map a property to an arithmetic_number. In regards to units of measure, the approach in ISO 15926-2:2003 is to classify the property_quantification, in other words a classification relation is used to map an instance of property_quantification to an instance of scale. The approach used here defines scale as an OWL:property.

```
 <owl:Class rdf:ID="temperature_quantity">
   <rdf:type rdf:resource="&lis;property_space"/>
   <rdfs:label>temperature_quantity</rdfs:label>
 </owl:Class>

<owl:FunctionalProperty rdf:ID="temperature_setpoint">
   <rdf:type rdf:resource="&lis;class_of_indirect_property"/>
   <rdfs:domain rdf:resource="&ctrl;controller"/>
   <rdfs:range rdf:resource="#temperature_quantity"/>
</owl:FunctionalProperty>

<lis:physical_object rdf:ID="TIC_01">
  <rdfs:comment>Temperature Controller TIC-01</rdfs:comment>
</lis:physical_object>

<owl:ObjectProperty rdf:ID="kelvin">
  <rdf:type rdf:resource="#scale"/>
  <rdfs:domain rdf:resource="#temperature_quantity"/>
  <rdfs:range rdf:resource="#real"/>
</owl:ObjectProperty>

<lis:physical_object rdf:ID="temporal_part_of_TIC_01_at_800K">
    <lis:temporal_whole_part.whole rdf:resource="#TIC_01"/>
    <temperature_setpoint>
      <rdf:Description>
        <rdf:type>
          <owl:Class rdf:about="#temperature_quantity"/>
```

```
      </rdf:type>
      <kelvin>
        <rdf:Description>
          <real>
            <content>
              <xsd:float rdf:value="800.0"/>
            </content>
          </real>
        </rdf:Description>
      </kelvin>
    </rdf:Description>
  </temperature_setpoint>
</lis:physical_object>
```

In this example, temperature_set_point is an instance of class_of_indirect-
_property defined so as to express that controllers can have a tempera-
ture_setpoint which accepts values of temperature but not pressure or any
other property_space. Note that Kelvin is defined in such a way that it would
be possible to detect inconsistencies in the units of measure of temperature
properties. The actual use of the scale Kelvin contains the value of 800 K,
meaning that controller TIC_01 had that value during a certain period of
time.

6.12 Time-reasoning

Temporal reasoning problems can be found in scheduling and planning systems, including problems such as minimizing assembly line slack time, project-
ing critical steps in a deployment plan to insure proper interaction between
them [23].

Let us assume that recipes are downloaded and the scheduling module in
the manufacturing system is requested to generate schedule alternatives with
a production start time between 8:30 and 10:00. In this type of situation, the
orderings of operations in the schedule must satisfy a number of constraints
including those imposed by the recipes (as a matter of fact the problem also
consists in finding which recipe is to be chosen).

Notice that the integration between the recipe and the scheduling tools
requires that the information representation of the ordering constraints in the
recipe to be consistent with the information representation of the same kind
of constraints in the schedule. This can be accomplished with the symbolic
time relations proposed by Allen shown in Figure 6.16:

precedes	for two activities A1 and A2, (precedes A1 A2) means that A1 ends before A2 begins
meets	for two arbitrary activities A1 and A2 (meets A1 A2) means that A2 begins at the time A1 ends.
overlaps	for two arbitrary activities A1 and A2 (overlaps A1 A2) means that A1 begins before A2 begins, and A1 ends before A2 ends
costarts	for two activities A1 and A2, (costarts A1 A2) means that A1 begins when A2 begins
cofinishes	for two arbitrary activities A1 and A2, (cofinishes A1 A2) means that A1 ends when A2 ends.
equals	for two arbitrary activities A1 and A2, (equals A1 A2) A1 and A2 will simultaneously end

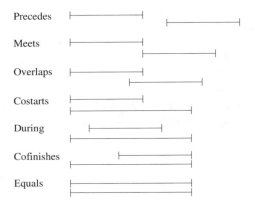

Fig. 6.16. Allen relations

6.13 Chapter Summary

¿From raw material procurement to product delivery, information systems have become ubiquitous assets in the manufacturing organizations. Information systems were firstly introduced at the factory floor and the tendency to automation continues to present times. ERP systems nowadays cover a wide range of functions intended to support the business. MRP systems were born to fit the gap between the factory-level control systems and the ERP . Unfortunately, investments on information technology tend to increase to an extent that the advantages may become overshadowed by the incurred costs. Worldwide enterprises spend considerable amounts of resources on data integration, which is associated to the ever-changing technologies, and the difficulties in integrating software from different vendors and legacy systems. To alleviate the situation, a variety of technologies have been developed that facilitate the task of integrating different applications. This chapter has discussed current integration technologies and ongoing research in this area.

References

1. IDC (2004) Worldwide data integration spending 2004–2008 Forecast. IDC Research

2. O'Leary D (2000) Enterprise resource planning systems systems, life cycle, electronic commerce and risk. The Cambridge University Press, UK

3. Harold ER, Means WS (2004) XML in a Nutshell. O'Reilly, CA

4. Bradshaw JM (1997) An Introduction to Software Agents. In JM Bradshaw (Ed.), Software Agents, 3–49, MIT Press, Cambridge, MA

5. Hendler J, McGuinness DL (2000) The DARPA agent markup language. IEEE Intelligent Systems, 15:67–73

6. FIPA (2005) FIPA specifications. [On-line] Available at: http://www.fipa.org/specifications/index.html

7. Shen W, Norrie D H (1999) Agent-based systems for intelligent manufacturing: a state-of-the-art survey. Knowledge and Information Systems 1:129–156

8. Tang HP, Wong TN (2005) Reactive multi-agent system for assembly cell control. Robotics and Computer-Integrated Manufacturing, 21:87–98

9. Fikes R, Jenkins J, and Gleb F (2003) JTP: A system architecture and component library for hybrid reasoning. Proceedings of the Seventh World Multiconference on Systemics, Cybernetics, and Informatics 2003; Orlando, Florida. July 27–30

10. Williams T J (1992) The Purdue enterprise reference model, a technical guide for CIM planning and implementation. Instrumentation, Systems and Automation Society. ISBN 1-55617-265-6

11. Chen D (2005) Enterprise–control system integration–an international standard. International Journal of Production Research 43:4335–4357

12. ISO 10303-11 (1994) Industrial automation systems and integration – product data representation and exchange – part 11: description methods: The EXPRESS language reference manual

13. Lacy LW (2005) OWL: representing information using the Web ontology language. Trafford Publishing, Victoria, Canada

14. Bechhofer S, van Harmelen F, Hendler J, Horrocks I, McGuinness DL, Patel-Schneider PF, and Stein LA (2004) OWL Web ontology language reference. http://www.w3.org/TR/owl-ref/

15. Finin T and Ding L (2006) Search engines for semantic Web knowledge. Proceedings of XTech 2006: Building Web 2.0, Amsterdam, May 16–19

16. Russell SJ and Norving P (1995) Artificial intelligence: a modern approach. Prentice Hall, Englewood Cliffs, NJ, USA

17. Niles I and Pease A (2001) Towards a standard upper ontology. 2nd International Conference on Formal Ontology in Information Systems (FOIS), Ogunquit, Maine, October 17–19

18. Sowa J (2000) Knowledge representation: logical, philosophical, and computational foundations. Brooks/Cole, Pacific Grove, CA

19. Gangemi A, Guarino N, Masolo C, Oltramari A, Schneider L (2002) Sweetening ontologies with DOLCE. Proceedings of EKAW 2002. Springer, Berlin

20. ISO 15926-2 (2003) ISO-15926:2003 Integration of lifecycle data for process plants including oil and gas production facilities: part 2 – data model

21. West M (2003) Replaceable parts: a four dimensional analysis. Proceedings of the Conference on Spatial Information Theory (COSIT). Ittingen, Switzerland, September 24–28

22. Gruber TR and Olsen GR (1994) An ontology for engineering mathematics. In J. Doyle, P. Torasso, and E. Sandewall (Eds.), Fourth International Conference on Principles of Knowledge Representation and Reasoning, Gustav Stresemann Institut, Bonn, Germany, Morgan Kaufmann

23. Stillman J, Arthur R, and Deitsch A (1993) Tachyon: a constraint-based temporal reasoning model and its implementation. SIGART Bulletin, 4:1–4

Summary

This book presented selected topics on recent developments in computing technologies for manufacturing systems. This covers three big areas, namely combinatorial optimization, fault diagnosis and monitoring and information systems to resolve difficult problems found in advanced manufacturing. These topics will be of interest to both mechanical and information engineers needing practical examples for the successful integration of scientific methodologies in manufacturing applications.

As an introductory remark, in Chap. 1, definitions, elements and concepts that configure the systems approach and characteristics of their functions were explained along with a transition of manufacturing systems. Then the content of the following chapters were featured briefly.

In Chap. 2, we focused on a variety of metaheuristic approaches that have emerged recently and are nowadays filtering as a practical optimization method by virtue of the rapid progress of both computers and computer science. They can also even cope with the combinatorial optimization readily. Due to these favorable properties, these methods are being widely applied to many difficult optimization problems often encountered in manufacturing.

Then, to solve various complicated and large-scale problems in a numerically effective manner, an illustrative formulation of a hybrid approach was presented in terms of the combination of traditional mathematical programming and recent metaheuristic optimization in a hierarchical manner.

To illustrate the effectiveness, three applications in manufacturing optimization were solved using each optimization method described here. Taking a logistic problem associated with supply chain management, a hybrid method was developed after decomposing the problem into a few appropriate sub-problems. Tabu search and the graph algorithm as an LP solver of the special class were applied to solve the resulting problems.

The second topic in this chapter concerned an injection sequencing problem under uncertainties associated with defective products. The result obtained from simulated annealing (SA) was shown to increase the efficiency of a mixed-model assembly line for small-lot-multi-kinds production.

Unlike the conventional simple model, the third topic concerned a realistic production scheduling involving multi-skilled human operators who can manipulate multiple types of resources such as machine tools, robots and so on. Such a general scheduling problem associated with the human tasks was formulated and solved by an empirical optimization method known as the dispatching rule in scheduling.

Since there exist more or less uncertain factors in mathematical models employed for optimization, we must pay careful attention to the uncertainties hidden in the optimization. As a new interest related to the recent development of metaheuristics, GA was applied to derive an insensitive solution against uncertain parameters.

Secondly, focusing on the logistic systems associated with supply chain management, the hybrid tabu search was applied to solve the problem under uncertain customer demand associated with the idea of flexibility analysis known in the field of process systems engineering. After classifying the decision variables as to whether they are soft (operation) or hard (design), it can derive a flexible logistic network against uncertainties just by adjusting the operation at the properly prescribed design.

Recently, multi-objective optimization has been used as a suitable decision aid supporting agile and flexible manufacturing under diversified customer demands. Chapter 3 focused on two different approaches to the multi-objective optimization.

The first one was associated with the multi-objective analysis in terms of extended applications of evolutionary algorithms (EA). Since the EA considers the multiple possible solutions simultaneously in the search, it can favorably generate a Pareto optimal solution set in a single run of the algorithm. Additionally, being insensitive to the feature of the Pareto front, it can deal with real world problems advantageously from the aspect of multi-objective analysis.

Then, a few multi-objective optimization methods in terms of soft computing (MOSC) associated with the methodology and applications in manufacturing systems were explained. Common to those methods, value function modeling methods using neural networks were presented.

Using the thus identified value function, a procedure of hybrid GA was extended to solve mixed-integer programs (MIP) under multi-objectives, even including qualitative objectives. Then, as the major interest of this chapter, the soft computing methods termed $MOON^2$ and $MOON^{2R}$ were presented with an extension to cases in the ill-posed decision environment.

At the early stages of product design, designers need to engage in model building as a step of the problem definition. Modeling of the value functions is also important in the designing task at the next stage. In such circumstances, the significance of integrating the modeling of both the system and the value function was emphasized as a key issue for competitive product development through multi-objective optimization.

To facilitate a wide application of MOSC in such a complex and global decision environment, a few applications ranging from a strategic planning to operational scheduling were demonstrated in the rest of this chapter.

First, the site location problem of a hazardous waste disposal site was solved by using the hybrid GA under the two objectives. The second topic concerned multi-objective scheduling optimization, and the effectiveness of MOSC using SA as an optimization method was illustrated for job shop scheduling. Thirdly, we illustrated a multi-objective design optimization taking a simple artificial product design and its extension for the integration of modeling and design optimization in terms of meta-modeling.

Though various models of associative memory have been studied recently, little attention has been paid to how to improve its capability for image processing or development of recognition in manufacturing systems.

From this aspect, in Chap. 4, taking CNNs for associative memory, a common design method was introduced by using singular value decomposition. Then, some new models such as the multi-valued output CNN and the multi-memory tables CNN were presented with applications to intelligent sensing and diagnosis.

Wavelet transform, which is a time-frequency method, has been receiving keen attention as a method for non-stationary signal analysis. It is classified into the continuous wavelet transform (CWT) and the discrete wavelets transform (DWT).

However, when CWT and DWT are used as a signal analysis method, some problems in the manufacturing systems arise. For example, in the case of CWT, it needs an enormous amount of computation and it is impossible to analyze the signals in real time. On the other hand, DWT cannot catch features of the signals exactly and has poor direction selection in the image. Chapter 5 focused on some useful methods to improve problems. The major methods are as follows:

Fast algorithm in the frequency domain for the CWT, the wavelet instantaneous correlation method by using the real signal mother wavelet for detecting and evaluating abnormal signals, the complex discrete wavelet transform by using the real-imaginary spline wavelet for improving the lack of translation invariance and poor direction selection.

Furthermore, these methods were applied to de-noising, abnormal detection, and image processing in manufacturing systems.

From raw material procurement to product delivery, information systems have become ubiquitous assets in manufacturing organizations. Chapter 6 discussed current integration technologies and ongoing research associated with the information systems from the following point of view. Unfortunately, investments in information technology tend to increase to the extent that the advantages may become overshadowed by the incurred costs. World-wide enterprises spend considerable amounts of resources on data integration, which is associated with the ever-changing technologies, and the difficulties in integrating software from different vendors and legacy systems. To alleviate the situation, a variety of technologies that facilitate the task of integrating different applications have been presented.

In the Appendices, after a brief introduction of IDEF0, traditional optimization methods of both single and multiple objectives were outlined for reference and in expectation of the emergence of a new type of hybrid approach. It covers the bases of optimization theory and algorithm as a whole. A pair-wise comparison quite similar to AHP is employed for the value function modeling of MOSC as well as feed forward neural networks. Hence, brief explanations were given for these components like AHP, BP, RBF networks, and ISM.

Generally speaking, it is not so difficult to apply a certain metaheuristic approach even to the complicated and large-scale problems in the real world. As a generic nature of the algorithm, however, the success will depend greatly also on the heuristic or trial and error tuning process. In addition to inventing a new method, automatic selection and/or combination of algorithms including hybridization and automatic tuning of algorithm parameters will be of special interest in future studies.

The cooperation of metaheuristic approaches with multi-objective optimization to construct the Pareto optimal solution set is becoming increasingly important. As a decision aid for supporting advanced manufacturing, however, its development should only be extended to several promising candidates for further consideration. On the other hand, MOSC can favorably satisfy such requirement. Since the major difficulty to engage in MOSC lies in the subjective judgment regarding preference, developing a user friendly interface amenable for this interaction is an important facet to facilitate these approaches.

Associative memory using CNN will be designed so as to correspond memory patterns to equilibrium points of the dynamics. For this purpose, a singular value decomposition method and new models of the multi-valued output CNN have been developed. However, since the network does not always converge efficiently to the memory patterns, the designed CNN cannot be guaranteed to be the most suitable. In order to resolve this problem, a new design method is expected with the development of the multi-valued output CNN having more than three output values. The pursuit of the possibility of CNN as the medium of associative memory is also left for future studies.

Today, wavelet transform is known as a popular signal analysis and image processing tool, and some new analysis methods such as the wavelet instantaneous correlation (WIC) method by using the real signal mother wavelet, and complex discrete wavelet transform (CDWT) by using the real-imaginary spline wavelet are being developed. By improving some properties such as the calculation speed of the WIC and perfect translation invariance in the CDWT, the wavelet transform will be applied more widely in manufacturing systems.

While much work has been done in manufacturing information systems, reconfiguration, proactive strategies, and knowledge integration are likely to become critical areas. Undoubtedly, system integration will be easier than it is today. We may find, for example, self-configuring applications and automatic integration approaches. Specific proactive strategies will result in multi-agent systems or their descendants with the ability to proactively carry out planning, operations execution and fault diagnosis in order to recover from abnormal situations.

Finally, this book gives relevant information for understanding technical details and assessing the research potential of computing technologies in manufacturing. It also provides a way of thinking toward sustainable manufacturing. Though facilitating sustainable technologies is a key issue for future directions associated with multi-disciplinary systems, it may be difficult to achieve this goal under global competition and also various conflicts between economical efficiency and greenness, industrial benefits and public welfare, *etc.* In view of this difficulty, it is essential to look at the subject as a whole and to establish a collaborative environment that can integrate each component readily. In order to achieve this, new methods and tools will be needed for orchestrating maintenance, real-time monitoring, simulation and optimization agents with planning, scheduling, design and control agents.

Appendix A

Introduction to IDEF0

IDEF0 (integrated definition for function modeling zero) is an activity modeling technique developed by the United States Air Force as a result of the Air Force's Integrated Computer Aided Manufacturing (ICAM) program. The IDEF0 activity modeling technique [1, 2], typically, aims at identifying and improving the flow of information within the enterprise, but it has been extended to cover any kind of process in which not only information but other resources are also involved. One use of the technique is to identify implicit knowledge about the nature of the business process, which can be used to improve the process itself (*e.g.*, [3, 4]). IDEF0 activity models can show which persons, teams or organizations participate in the same activity and the existing software tools that support such activity. For example, this helps identify which computing technology is necessary to perform a specific activity. Activity modeling shows the information that is used or produced by an activity. Consequently, data requirements can be identified for producing an information model and ontologies such as those described in Chap. 6.

IDEF0 activity models are developed in hierarchical levels. It is possible, therefore, to start with a high-level view of the process that is consistent with global goals, and then decompose it into layers of increasing details. A rectangular box graphically represents each activity with four arrows reading clockwise around the box as shown in the upper part of Figure A.1. These arrows are also referred to as ICOM (inputs, constraints or controls, outputs and mechanisms). Input is the information, material or energy used to produce the output of an activity. The input is going to be acted upon or transformed to produce the output. Constraint or control is the information, material or energy that constrains and regulates an activity. Output is the information, material or energy produced by or resulting from the activity. Mechanism represents the resources such as people, equipment, or software tools that perform an activity. After all, the relation between input and output represents what is done through the activity, while control describes why it is done, and the mechanism by which it is done.

An IDEF0 diagram is composed of the following:

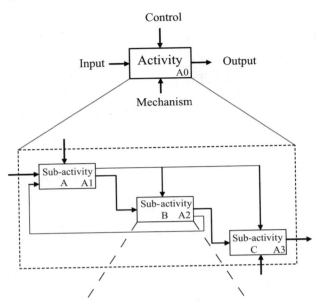

Fig. A.1. A basic and extended structures of IDEF0

1. A top level diagram that illustrates the highest level activity and its ICOMs.
2. Decomposition diagrams, which represent refinements of an activity by showing its lower level activities, their ICOMs, and how activities in the diagram relate to each other.
3. A glossary that defines the terms or labels used on the diagrams as well as natural language descriptions of the entire diagram.

Activities are named by using active verbs in the present tense, such as "design product," "simulate process," "evaluate plant," *etc.* Also all decomposed activities have node identifiers that begin with a capital letter and numbers that show the relation between a parent box and its child diagrams. The A0 top level activity is broken down into the next level of activities with node numbers A1, A2, A3, *etc.*, which in turn are broken down and at the next level labeled A11, A12, A13, *etc.* In modeling activities, it is important to keep in mind that they will define the tasks that cross-functional teams and tools will perform. Because different persons may develop different activity models, it is important to define requirements and context at the outset of the process improving process. From this aspect, its simple modeling rules are very helpful for easy application, and its hierarchical representation is suitable to grasp a whole idea quickly without dwelling on the precise details too much.

This hierarchical activity modeling technique endows us with the following favorable properties suitable for the activity modeling in manufacturing.

1. Explicit description about information in terms of the control and the mechanism in each activity is helpful to set up some sub-goals for the evaluation.
2. We can use appropriate commercial software having various links with simulation tools to evaluate certain important features of business process virtually.
3. Since the business process belongs to a cooperative work of multi-disciplinary nature, the IDEF0 provides a good environment to share common recognition among them.
4. Having a structure to facilitate modular design, the IDEF0 is easy to modify and/or correct the standard model corresponding to the particular concerns.

References

1. Marca DA, McGowan CL (1993) IDEF0/SADT business process and enterprise modeling. Eclectic Solutions Corporation, San Diego
2. Colquhoun GJ, Baines RW, Crossley R (1993) A state of the art review of IDEF0. International Journal of Computer Integrated Manufacturing, 6:252–264
3. Colquhoun GJ, Baines RW (1991) A generic IDEF0 model of process planning. International Journal of Production Research, 11:2239–2257
4. OSullivan D (1991) Project management in manufacturing using IDEF0. International Journal of Project Management, 9:162–168

Appendix B

The Basis of Optimization Under a Single Objective

B.1 Introduction

Let us review briefly traditional optimization methods under a single-objective function or usual optimization methods in mathematical programming (MP) . Optimization problems are classified depending on their properties as follows:

- Form of equations
  1. Linear programming problem (LP)
  2. Quadratic programming problem (QP)
  3. Nonlinear programming problem (NLP)
- Property of decision variables
  1. (All) integer programming problem (IP)
  2. Mixed-integer programming problem (MIP)
  3. (All) zero-one programming problem
  4. Mixed-zero-one programming problem
- Number of objective functions
  1. Single-objective problem
  2. Multi-objective problem
- Concern with uncertainty
  1. Deterministic programming problem
  2. Stochastic programming problem
     - expectation-based optimization
     - chance-constraint optimization
  3. Fuzzy programming problem
- Size of the problem
  1. Large-scale problem
  2. Medium-scale problem
  3. Small-scale problem

Since a description covering all of these[1] is beyond the scope of this book, only an essence of several methods that are still important today will be explained to give a basis for understanding the contents of the book.

B.2 Linear Programming and Some Remarks on Its Advances

We start with introducing a linear program or a linear programming problem (LP) that can be expressed in standard form as follows:

$$[Problem] \quad \min \quad z = c^T x \quad \text{subject to} \quad \begin{cases} Ax = b \\ x \geq 0 \end{cases},$$

where x is an n-dimensional vector of decision variables, and A (($m \times n$)-dimension) and b (m-dimension) are a coefficient matrix and a vector of the constraints, respectively. Moreover, c (n-dimension) is a coefficient vector of objective function, and T denotes the transpose of a vector and/or a matrix. All these dimensions must be consistent for matrix and/or vector computations. Matrix A generally has more columns than rows, $i.e.$, ($n > m$). Hence the simultaneous equation $Ax = b$ is under determined, and this allows choosing x to minimize $c^T x$. Assuming every equation involved in the standard form is not redundant, or the rank of matrix A is equal to the number of constraints m, let us divide the vector of decision variables into two sub-sets representing an m-dimensional basic variable vector x_B and a non-basic variable vector composed of the remaining variables x_{NB}. Then, rewrite the original objective function and constraints accordingly as follows:

$$z = c^T x = (c_B^T, \ c_{NB}^T) \begin{pmatrix} x_B \\ x_{NB} \end{pmatrix} = c_B^T x_B + c_{NB}^T x_{NB},$$

$$Ax = [B, \ A_{NB}] \begin{pmatrix} x_B \\ x_{NB} \end{pmatrix} = b,$$

where c_B and B denote a sub-vector and a sub-matrix corresponding to x_B, respectively. It should be noticed here that B becomes a square matrix. On the other hand, c_{NB} and A_{NB} are a sub-vector and a sub-matrix for x_{NB}. For an appropriately chosen x_B, it is supposed that the matrix B is regular or it has an inverse matrix B^{-1}. Then we have the following equations:

$$x_B = B^{-1}(b - A_{NB}x_{NB})$$
$$= B^{-1}b - B^{-1}A_{NB}x_{NB}, \tag{B.1}$$

---

[1] Refer to other textbooks [1, 2, 3, 4], for examples.

$$z = c_B^T B^{-1} b + (c_{NB}^T - c_B^T B^{-1} A_{NB}) x_{NB}. \tag{B.2}$$

Since the numbers of solution are finite, say at most $_nC_m$, we can find the global optimal solution with a finite computation load by simply enumerating all possible solutions. However, such a load expands rapidly as n and/or m become large. The solution forcing $x_{NB} = 0$ or $x^T = (x_B^T, 0^T)$ is called a basic solution. Any feasible solution and its objective value can be obtained from the solution of the following linear simultaneous equations:

$$\begin{bmatrix} B & 0 \\ -c_B^T & 1 \end{bmatrix} \begin{pmatrix} x_B \\ z \end{pmatrix} = \begin{pmatrix} b \\ 0 \end{pmatrix}.$$

As long as there is a solution, the above equation can be solved as Equation B.4 by noticing the following formula:

$$\hat{B}^{-1} = \begin{bmatrix} B & 0 \\ -c_B^T & 1 \end{bmatrix}^{-1} = \begin{bmatrix} B^{-1} & 0 \\ c_B^T B^{-1} & 1 \end{bmatrix}, \tag{B.3}$$

$$\begin{pmatrix} x_B \\ z \end{pmatrix} = \hat{B}^{-1} \begin{pmatrix} b \\ 0 \end{pmatrix} = \begin{pmatrix} B^{-1} b \\ c_B^T B^{-1} b \end{pmatrix}. \tag{B.4}$$

This expression is equivalent to the results obtained from Equations B.1 and B.2 by letting x_{NB} equal zero. From the discussions so far, it is easy to understand that the particular basic solution becomes optimal when the following conditions hold:

$$\begin{cases} B^{-1} b \geq 0 \\ c_{NB}^T - c_B^T B^{-1} A_{NB} \geq 0^T \end{cases}.$$

These equations are known as the feasibility and the optimality conditions, respectively. Though these conditions provide necessary and sufficient conditions for the optimality, they say nothing about a procedure how to obtain the optimal solution in practice.

The simplex method developed by Dantzig [5] more than 40 years ago has been popularly known as the most effective method for solving linear programming problem for a long time. It takes an iterative procedure by noticing that the basic solutions represent extreme points of the feasible region. Then the simplex method searches from one extreme point to another one along the edges of the boundary of the feasible region toward the optimal point successively.

By introducing slack variables and artificial variables, its solution procedure begins with transforming the original Problem B.5 into the standard form like Problem B.6,

$$[Problem] \quad \min \quad z = c^T x \quad \text{subject to} \quad \begin{cases} A_1 x \le b_1 \\ A_2 x = b_2 \\ A_3 x \ge b_3 \end{cases}, \quad (B.5)$$

$$[Problem] \quad \min \quad z = c^T x \quad \text{subject to} \quad \begin{cases} A_1 x + s_1 = b_1 \\ A_2 x + w_2 = b_2 \\ A_3 x - s_3 + w_3 = b_3 \end{cases}, \quad (B.6)$$

where s_1 and s_3 denote slack variable vectors, and w_2 and w_3 artificial variable vectors. Rearranging this like

$$[Problem] \quad \min \quad z = (c^T, \ 0^T, \ 0^T, \ 0^T, \ 0^T) \begin{pmatrix} x \\ s_3 \\ s_1 \\ w_2 \\ w_3 \end{pmatrix},$$

$$\text{subject to} \quad \begin{bmatrix} A_1 & 0 & I_1 & 0 & 0 \\ A_2 & 0 & 0 & I_2 & 0 \\ A_3 & -I & 0 & 0 & I_3 \end{bmatrix} \begin{pmatrix} x \\ s_3 \\ s_1 \\ w_2 \\ w_3 \end{pmatrix} = \begin{pmatrix} b_1 \\ b_2 \\ b_3 \end{pmatrix},$$

we can immediately select s_1, w_2, and w_3 as the basic variable vectors. Following the foregoing notations, the simplex method describes this status as the following simplex tableau:

$$\begin{bmatrix} A_{NB} & I & b \\ -c_{NB}^T & 0^T & 0 \end{bmatrix}. \quad (B.7)$$

Here, the following correspondence should be noticed:

$$A_{NB} = \begin{bmatrix} A_1, & 0 \\ A_2, & 0 \\ A_3, & -I \end{bmatrix}, \quad I = \begin{bmatrix} I_1 & 0 & 0 \\ 0 & I_2 & 0 \\ 0 & 0 & I_3 \end{bmatrix}, \quad b = \begin{pmatrix} b_1 \\ b_2 \\ b_3 \end{pmatrix},$$

$$c_{NB}^T = (c^T, \ 0^T), \quad c_B = 0,$$

$$x_{NB}^T = [x^T, \ s_3^T], \quad x_B^T = (s_1^T, \ w_2^T, \ w_3^T).$$

Since such a solution that $s_1 = b_1$, $w_2 = b_2$, $w_3 = b_3$, and $x = s_3 = 0$ is apparently neither optimal nor feasible, we need to move toward the optimal solution while recovering the infeasibility in the following steps.

Before considering this, it is meaningful to review the procedure known as pivoting in the simplex method. It is an operation to replace a basic variable with a non-basic variable in the current solution to update the basic solution.

This can be carried out by multiplying the matrix expressed in Equation B.3 from the left-hand side to the matrix of Equation B.7:

$$\begin{bmatrix} B^{-1} & 0 \\ c_B^T B^{-1} & 1 \end{bmatrix} \begin{bmatrix} A_{NB} & I & b \\ -c_{NB}^T & 0^T & 0 \end{bmatrix} = \begin{bmatrix} B^{-1} A_{NB} & B^{-1} & B^{-1}b \\ c_B^T B^{-1} A_{NB} - c_{NB}^T & c_B^T B^{-1} & c_B^T B^{-1}b \end{bmatrix},$$

As long as the condition $c_B^T B^{-1} A_{BN} - c_{NB}^T > 0$ holds, we can improve the current solution by continuing the pivoting. Usually, the non-basic variable with the greatest value of this term, say s, will be selected first as a new basic variable. Then according to this choice, will be withdrawn such a basic variable that becomes critical to keep the feasibility condition $B^{-1}b \geq 0$, i.e., $\min_{j \in I_B} \hat{b}_j / a_{js}$, (for $a_{js} > 0$). Here I_B is an index set denoting the basic variables, and a_{js}, (j, s)-element of the tableau, and $\hat{b}_j$ the current value of the j-th basic variable. Substituting $c_B^T B^{-1} = \pi^T$ (simplex multiplier), the above matrix can be rewritten compactly as follows:

$$\begin{bmatrix} B^{-1} A_{NB} & B^{-1} & B^{-1}b \\ \pi^T A_{NB} - c_{NB}^T & \pi^T & \pi^T b \end{bmatrix}.$$

Now let us go back to the problem of how to sweep out the artificial variables that appear by transforming the problem into the standard form. We can obtain a feasible solution if and only if we sweep out every artificial variable from the basic variables. To work with this problem, there exist two major methods, known as the two-phase method and the penalty function method. The two-phase method tries to recover from the infeasibility first, and then turns to optimization. On the other hand, the penalty function method will consider only the optimal condition. Instead, it urges the artificial variables to leave the basic solutions as soon as possible, and restricts them from coming back to the basic solutions once they have left the basic variables.

In the two-phase method, an auxiliary linear programming problem is solved first under the following objective function:

$$[Problem] \quad \min \quad v = \sum_i w_{2i} + \sum_i w_{3i}.$$

If and only if every artificial variable becomes zero, does the optimal value of this objective function also become zero. This is equivalent to saying that there exists a feasible solution in the present problem since every artificial variable has been swept out or turned to the non-basic variables at this stage. Now we can continue the same procedure under the original objective function until the optimality condition has been satisfied.

On the other hand, the penalty function method will modify the original objective function by augmenting penalty terms as follows:

$$[Problem] \quad \min \quad z' = \sum_i c_i x_i + \left(M_2 \sum_i w_{2i} + M_3 \sum_i w_{3i} \right).$$

Due to the large values of penalty coefficients M_2 and M_3, the artificial variables are likely to leave the basic variables and be restricted to the basic variables again once they have left.

There are many interesting findings to be noted regarding the simplex method and LP, for examples, the graphical solution method and a geometric understanding of the search process; the revised simplex method to improve the solution efficiency; degeneracy of basic variables; the dual problem and its relation to the primal problem; dual simplex method, sensitivity analysis, *etc.*

Recently, a new algorithm known as the interior-point method [6] has been shown especially efficient for solving very large problems. By noticing that such problem has a very sparse coefficient matrix, these methods are developed based on the techniques from nonlinear programming. Though the simplex method visits the extreme points one after another along with the ridges of the admissible region, the interior-point methods search the inside of the feasible region while improving a series of tentative solutions.

The successive linear programming and separable linear programming are extended applications of the ordinal method. In addition to these mathematically interesting aspects, the importance of LP is due to the existence of good general-purpose software for finding the optimal solution (not only commercial but also free software is available from the Web [7]).

As a variant of LP, integer programs (IP) requires all variables to take integer values, and mixed-integer programming (MIP) requires some of the variables to take integer values and others real values. As a special class of these programs, zero-one IP or zero-one MIP, which restrict their integer variables only to zero or one, are widely applicable since manifold combinatorial and logical conditions can be modeled through zero-one variables. These classes of programs often have the advantage of being more realistic than LPs, but the disadvantage of being much harder to solve due to the combinatorial nature of the solution. The most widely available general-purpose technique for solving these problems is a procedure called "branch-and-bound (B & B) method" [8]. It tries to search the optimal solution by deploying a tree of potential solutions derived from the related LP relaxation problem that allows integer variables to take real numbers.

In the context of LP, there are certain models whose solution always turns out to be integer when every coefficient of the problem is integer. This class is known as the network linear programming problem [9], and make it unnecessary to deal with the problem as difficult as MIP or IP. Moreover, it can be solved 10 to 100 times faster than general linear programs by using specialized routines of the simplex method. It tries to minimize the total cost of flows along all arcs of the network subject to conservation of flow at each node, and upper and/or lower bounds on the flow along each arc.

The transportation problem is an even more special case in which the network is bipartite: all arcs run from nodes in one subset to the nodes in a disjoint subset. In the minimum cost flow problem in Sect. 2.4.1, a network is composed of a collection of nodes (locations) and arcs (routes) connecting selected pairs of nodes. Arcs carry a physical or conceptual flow, and may be directed (one-way) or undirected (two-way). Some nodes become sources (permitting flow to enter the network) or sinks (permitting flow to leave).

A variety of other well-known network problems such as shortest path problems solved by Dijkstra's method in Sect. 2.5.2, maximum flow problems, and certain assignment problems can also be modeled and solved like the network linear programs.

Industries have made use of LP and its extensions for modeling a variety of problems in planning, routing, scheduling, assignment, and design. In future, they will continue to be valuable for problem-solving including transportation, energy, telecommunications, and manufacturing in many fields.

B.3   Non-linear Programs

Non-linear programs or the non-linear programming problem (NLP) has a more general form regarding the objective function and constraints, and is described as follows:

$$[Problem] \quad \min \ f(x) \ \text{ subject to } \ \begin{cases} g_i(x) \geq 0, \ (i = 1, \ \ldots, m_1) \\ h_j(x) = 0, \ (j = m_1 + 1, \ \ldots, m) \end{cases} ,$$

where x denotes an n-dimensional decision variable vector. Such a problem that all the constraints $g(x)$ and $h(x)$ are linear is called linearly constrained optimization, and if the objective function is quadratic, it is known as quadratic programming (QP) . Another special case where there are no constraints at all is called unconstrained optimization.

Most of the conventional methods of NLP encounter some problems associated with the local optimum that will satisfy the requirements only on the derivatives of the functions. In contrast, real world problems often have an objective function with multiple peaks, and pose difficulties for an algorithm that needs to move from a peak to a peak until attaining at the highest one. Algorithms that can overcome this difficulty are termed global optimization methods, and most recent metaheuristic approaches mentioned in the main text have some advantages on this point.

Since any equality constraint can be described by a pair of inequality constraints ($h(x) = 0$ is equivalent to the conditions $h(x) \geq 0$ and $h(x) \leq 0$), it is enough to consider the problem only under the inequality constraints. Without losing generality, therefore, let us consider the following problem:

[*Problem*]   min $f(x)$ subject to $g(x) \geq 0$.

Under mild mathematical conditions, the Karush-Kuhn–Tucker conditions give necessary conditions for this problem. These conditions also become sufficient under a certain condition regarding convexity as mentioned below. Let us start by giving the Lagrange function as follows:

$$L(x,\ \lambda) = f(x) - \lambda^T g(x),$$

where λ is a Lagrange multiplier vector. Thus by transforming the constrained problem into an unconstrained one superficially in terms of Lagrange multipliers, the necessary conditions for the optimality will refer to the stationary condition of the Lagrange function. Here x^* becomes a stationary point of function $f(x)$ if the following extreme condition is satisfied:

$$(\partial f/\partial x)_{x^*} = \bigtriangledown f(x^*) = 0^T. \tag{B.8}$$

Moreover, the sufficient conditions for a minimal extremum are given by

$$\bigtriangledown f(x^*) = 0^T,$$
$$[\partial(\partial f/\partial x)^T/\partial x]_{x^*} = \bigtriangledown^2 f(x^*) \ \text{(Hesse matrix) is positive definite.}$$

Here, we call matrix A positive definite if $d^T A d > 0$ holds for an arbitrary $d(\neq 0) \in \mathbb{R}^n$, and positive semi-definite if $d^T A d \geq 0$. A so-called saddle point locates on the point where it is neither negative nor positive definite. Moreover, function $f(x)$ $(-f(x))$ is termed a convex (concave) function when the following relation holds for an arbitrary $\alpha, (0 \leq \alpha \leq 1)$ and $x^1, x^2 \in \mathbb{R}^n$:

$$f(\alpha x^1 + (1 - \alpha)x^2) \leq \alpha f(x^1) + (1 - \alpha)f(x^2).$$

Finally, the stationary conditions of the Lagrange function making x^* a local optimum point for the constrained problem are known as the following Karush–Kuhn–Tucker (KKT) conditions:

$$\begin{cases} \bigtriangledown_x L(x^*,\ \lambda^*) = (\partial f/\partial x)_{x^*} - \lambda^{*T}(\partial g/\partial x)_{x^*} = 0^T \\ \bigtriangledown_\lambda L(x^*,\ \lambda^*)^T = g(x^*) \geq 0 \\ \lambda^{*T} g(x^*) = 0 \\ \lambda^* \geq 0 \end{cases}.$$

When $f(x)$ is a convex function and the feasible region prescribed by $g(x) \geq 0$ is a convex set, the above formulas also give the sufficient conditions. Here, a convex set is defined as a set satisfying the conditions that when both x^1 and x^2 are contained in a certain set S, $\alpha x^1 + (1 - \alpha)x^2$ is also a member of S for an arbitrary α $(0 \leq \alpha \leq 1)$.

The KKT conditions that neglect $g(x)$ and λ accordingly are equivalent to those of the unconstrained problem, or simply the extreme condition shown in Equation B.8. The linearly constrained problem guarantees the convexity of the feasible region, and QP has a concave objective function and a convex feasible region.

Golden section search and the Fibonacci method are popular algorithms for deriving the optimal solution numerically for the unconstrained problem with a scalar decision variable. Though they seem to be too simple to deal with real world applications, they are conveniently used as a subordinate routine of various algorithms. For example, many gradient methods require finding the step size to the prescribed search direction per iteration. Since this refers to a scalar unconstrained optimization, these methods can serve conveniently for such a search.

Besides these scalar optimization methods, a variety of pattern search algorithms have been proposed for vector optimization so far, *e.g.*, the Hooke–Jeeves method [10], the Rosenbrock method [11], *etc.* Among them, here we cite only the simplex method for unconstrained problems, and the complex method for constrained ones. These methods can have some connection to the relevant metaheuristic methods. It is promising to use these methods in a hybrid manner as a generating technique for initial solutions, an algorithm for the local search and a refining procedure at the end of search.

The simplex method[2] is a common numerical method for minimizing the unconstrained problem in an n-dimensional space. The preliminary idea was originally proposed by Himsworth, Spendley and Hex, and then extended by Nelder and Mead [17]. In this method, a geometric figure termed simplex plays a major role in the algorithm. It is a polytope of $n+1$ vertices in n-dimensional space, and has a structure that can easily produce a new simplex by taking reflection of the specific vertex with respect to the hyper-plane spanned by the remaining vertices. In addition, the reflection to the worst vertex may give a promisingly better solution. Relying on these properties of the simplex, the algorithm is deployed only by three operations mentioned below. Beforehand, let us specify the following vertices for the minimization problem:

1. x^h is a vertex such that $f(x^h) = \max_i \{ f(x^i), \ i = 1, 2, \ldots, n+1 \}$.
2. x^s is a vertex such that $f(x^s) = \max_i \{ f(x^i), i = 1, 2, \ldots, n+1, \ i \neq h \}$.
3. x^l is a vertex such that $f(x^l) = \min_i \{ f(x^i), \ i = 1, 2, \ldots, n+1 \}$.
4. x^G is the center of gravity of the simplex except for $i \neq h$, *i.e.*, $x^G = \sum_{i=1, \ i \neq h}^{n+1} x^i / n$.

By applying the following operations depending on the case, a new vertex will be generated in turn (see also Figure B.1):

- Reflection: $x^r = (1 + \alpha)x^G - \alpha x^h$,
  where $\alpha (> 0)$ is a constant and a rate of distance $(x^r - x^G)$ to $(x^h - x^G)$. This is the basic operation of this method.

---

[2] The name is same as a method of LP.

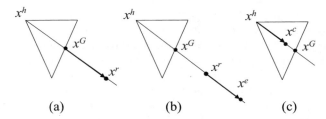

Fig. B.1. Basic operations of the simplex method: (a) Reflection, (b) expansion, (c) contraction

- Expansion: $x^e = (1 - \gamma)x^G + \gamma x^r$,
  where $\gamma(> 1)$ is a constant and a rate of distance $(x^e - x^G)$ to $(x^r - x^G)$. This operation takes place when the further improvement is promising beyond x^r in the direction $(x^r - x^G)$.
- Contraction: $x^c = (1 - \beta)x^G + \beta x^h$,
  where β (< 1) is a constant and a rate of distance $(x^c - x^G)$ to $(x^h - x^G)$. This operation shrinks the simplex when x^r fails. Generally, this will frequently appear at the end of search.

The algorithm is outlined below.

Step 1: Let $t = 0$. Generate the initial vertices, and specify x^h, x^s, x^l among them by evaluating each objective function, and calculate x^G.

Step 2: Apply the reflection to obtain x^r.

Step 3: Produce a new simplex from one of the following operations.
  3-1: If $f(x^l) \leq f(x^r) \leq f(x^s)$, replace x^h with x^r.
  3-2: If $f(x^r) < f(x^l)$, further improvement is expectable toward $x^r - x^G$. Apply the expansion, and see whether $f(x^e) < f(x^r)$ or not. If it is, replace x^h with x^e. Otherwise go back to x^r, and replace x^h with x^r.
  3-3: If $f(x^s) \leq f(x^r) < f(x^h)$, apply the contraction after replacing x^h with x^r. In the case of $f(x^r) \geq f(x^h)$, contract without such substitution. After either of these operations, if $f(x^c) < f(x^h)$, replace x^h with x^c. Otherwise shrink the simplex entirely toward x^l, i.e., $x^i := (x^i + x^l)/2, (i = 1, 2, \ldots, n + 1, i \neq l)$.

Step 4: Examine the stopping condition. If satisfied, stop. Otherwise, go back to Step 2.

Similar to most conventional multi-dimensional optimization algorithms, this occasionally gets stuck at a local optimum. The common approach to resolve this problem is to restart the algorithm with a new simplex starting at the current best value.

This method is also known as the flexible polyhedron method. Relating to such a name, we can compare this method to one of the recent metaheuristic methods if we view the simplex as a life like ameba. According to a certain stimulus, it will stretch and/or shrink its tentacle to the target, *e.g.*, food, chemicals, *etc.*

Many variants of the method exist depending on the nature of actual problem being solved. For example, an easy extension for the constrained problem is to move the new vertex x' on its boundary x^b when it violates the constraints. In the case of linearly constrained problem ($a_i^T x \leq b_i$, ($i = 1, 2, \ldots, m$)), the boundary point is easily calculated by

$$x^b = x^G + \lambda^*(x^G - x^h),$$

where λ^* is a constant decided from

$$\min_{i \in I_{\text{vio}}} \lambda_i = \frac{b_i - a_i^T x^G}{a_i^T(x^G - x^h)}, \quad I_{\text{vio}} = \{i \mid a_i^T x' > b_i\}.$$

The complex method developed by M.J. Box [13] is available for the constrained optimization problem subject to the constraints shown below,

$$G^i \leq x \leq H^i \ (i = 1, 2, \ldots, m),$$

where the upper and lower constraints H^i and G^i are either constants or nonlinear functions of decision variables. The feasible region subject to such constraints is assumed to be a convex set and there exists at least one feasible solution in it.

Since the simplex method uses ($n + 1$) vertices, its shape tends to become flat near the boundary of the constraints as a result of pulling back the violated vertex. Consequently, the vertex is likely to become trapped in a small subspace adhering to the hyper-plane parallel to the boundary. In contrast, the complex method employs a complex composed of k ($> n+1$) vertices to avoid such flattening. Its procedure is outlined below[3].

Step 1: An initial complex is generated by a feasible starting vertex and $k-1$ additional vertices derived from $x^i = G^i + r_i(H^i - G^i)$, ($i = 1, \ldots, k-1$) where r_i is a random number between 0 and 1.

Step 2: The generated vertices must satisfy both the explicit and implicit constraints. If at any time the explicit constraints are violated, the vertex is moved a small distance δ inside the boundary of the violated constraint. If an implicit constraint is violated, the vertex point is moved a half of the distance to the centers of gravity of the remaining vertices, *i.e.*, $x_{\text{new}}^j := (x_{\text{old}}^j + x^G)/2$, where the center of gravity of the remaining vertices x^G is calculated by

---

[3] Box recommends values of $\alpha = 1.3$ and $k = 2n$.

$$x^G = \frac{1}{k-1}\left(\sum_{j=1}^{k-1} x^j - x_{\mathrm{old}}^j\right).$$

This process is repeated until all the implicit constraints are satisfied. Then the objective function is evaluated at each vertex.

Step 3 (Over-reflection): The vertex having the highest value is replaced with a vertex x^O calculated by the following equation (see also Figure B.2):

$$x^O = x^G + \alpha(x^G - x^h).$$

Step 4: If x^O might give the highest value on consecutive trials, it is moved a half of the distance to the center of gravity of the remaining points.

Step 5: Thus resulting vertex is checked as to whether it satisfies all constraints or not. If it violates any constraints, adjust it as before.

Step 6: Examine the stopping condition. If satisfied, stop. Otherwise go back to Step 3.

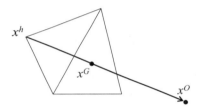

Fig. B.2. Over-reflection of the complex method

Both methods mentioned above are called "direct search" since their algorithms use only the evaluated value of the objective function. This is the merit of the direct search since the other methods require some information on the derivatives of function, which is not always easy to calculate in real world problems.

In spite of this, various gradient methods are very popular for solving both unconstrained problems and constrained ones. In the latter case, though a few methods try to calculate the gradient through projection on the constrained boundaries, some penalty function methods are usually employed to consider the constraints conveniently.

The Newton–Raphson method is a straightforward extension of the Newton method, which is a method to solve the algebraic equation numerically. Since the necessary conditions for optimality are given by an algebraic equation derived from first-order differentiation (*e.g.*, Equation B.8), application of the Newton method to the optimization needs second-order differentiation eventually. It is known that the convergence is rapid, but the computational load is considerable.

As one of the most effective methods, the sequential quadratic programming method (SQP) has been widely applied recently. It is an iterative solution method that updates the tentative solution of QP successively. Owing to the favorable properties of QP for solving problems in its class, SQP provides a fast convergence with a moderate amount of computation.

References

1. Chong EKP, Zak SH (2001) An introduction to optimization (2nd ed.). Wiley, New York
2. Conn AR, Gould NIM, Toint PL (1992) Lancelot: a FORTRAN package for large-scale nonlinear optimization (release A). Springer, Berlin
3. Polak E (1997) Optimization: algorithms and consistent approximations. Springer, New York
4. Taha HA (2003) Operations research: an introduction (7th ed.). Prentice Hall, Upper Saddle River
5. Dantzig G.B (1963) Linear programming and extensions. Princeton University Press, Princeton
6. Karmarkar N (1984) A new polynomial-time algorithm for linear programming. Combinatorica, 4:373–395
7. http://groups.yahoo.com/group/lp_solve/
8. Land AH, Doig AG (1960) An automatic method for solving discrete programming problems. Econometrica, 28:497–520
9. Hadley G (1962) Linear programming. Addison–Wesley, Reading, MA
10. Hooke R, Jeeves TA (1961) Direct search solution of numerical and statistical problems. Journal of the Association for Compututing Machinery, 8:212–229
11. Rosenbrock P (1993) An automatic method for finding the greatest or least value of a function. Computer Journal, 3:175–184
12. Nelder JA, Mead R (1965) Simplex method for functional minimization. Computer Journal, 7:308–313
13. Box MJ (1965) A new method of constrained optimization and a comparison with other methods. Computer Journal, 8:42–52

Appendix C

The Basis of Optimization Under Multiple Objectives

C.1 Binary Relations and Preference Order

In what follows, some mathematical basis of multi-objective optimization (MOP) will be summarized while leaving more detailed explanation to other textbooks [1, 2, 3, 4, 5, 6, 7].

A binary relation $R(X, Y)$ is a subset of the Cartesian product of the vector set X and Y having the following properties.

[**Definition 1**] A binary relation R on X is

1. reflexive if xRx for every $x \in X$.
2. asymmetric if $xRy \rightarrow$ not yRx for every $x, y \in X$.
3. anti-asymmetric if xRy and $yRx \rightarrow x = y$ for every $x, y \in X$.
4. transitive if xRy and $yRz \rightarrow xRz$ for every $x, y, z \in X$.
5. connected if xRy or yRx (possibly both) for every $x, y \in X$.

When a set of alternatives is denoted as A, a mapping from A to the consequence set is described such that $X(A) : A \rightarrow X$. Since it is adequate for the decision maker (DM) to rank his/her preference over the alternatives in the consequence space, the following concerns should be addressed on this set.

The binary relation $\preceq$ on $X(A)$ or X_A will be called the preference relation of the DM and classified as follows (read $x \preceq y$ as y is preferred or indifferent to x).

[**Definition 2**] A binary relation $\preceq$ on a set X_A is

1. weak order $\leftrightarrow \preceq$ on X_A is connected and transitive.
2. strict order $\leftrightarrow \preceq$ on X_A is anti-symmetric weak order.
3. partial order $\leftrightarrow \preceq$ on X_A is reflexive and transitive.

In terms of $\preceq$, two additional relations termed indifference $\sim$ and strict preference $\prec$ are defined on X_A as follows.

[Definition 3] A binary relation $\sim$ on X_A is an indifference if $x \sim y \leftrightarrow$ $(x \preceq y,\ y \preceq x)$ for every $x, y \in X$.

[Definition 4] A binary relation $\prec$ on X_A is a strict preference if $x \prec$ $y \leftrightarrow (x \preceq y,\ \text{not } y \preceq x)$ for every $x, y \in X$.

Now we will present some well-known properties without proof below.

[Theorem 1] If $\preceq$ on X_A is a weak order, then

1. exactly one of $x \prec y$, $y \prec x$, $x \sim y$ holds for each $x, y \in X_A$.
2. $\prec$ is transitive, $\sim$ is an equivalence (reflexive, symmetric and transitive).
3. $(x \prec y,\ y \sim z), \to x \prec z$ and $(x \sim y,\ y \prec z) \to x \prec z$.
4. $\preceq'$ on the set of equivalence classes of X_A under $\sim$, $X_A/\sim$ is a strict order where $\preceq'$ is defined such that $a \preceq' b \leftrightarrow x \preceq y$ for every $a, b \in X_A/\sim$ and some $x \in a$ and $y \in b$.

From the above theorem, it is predictable that there is a real-valued function that preserves the order on $X_A/\sim$. In fact, such existence is proven by the following theorem.

[Theorem 2] If $\preceq$ on X_A is a weak order and $X_A/\sim$ is countable, then there is a real-valued function $u(x)$ on X_A such that $x \preceq y \leftrightarrow u(x) \leq u(y)$ for every $x, y \in X_A$.

The above function $u(x)$ is termed a utility function, and is known to be unique in the sense that the preference order is preserved regarding arbitrary monotonic increasing transformations. Therefore, if the explicit form of the utility function is known, multi-objective optimization is reduced to a usual single-objective optimization of $u(x)$.

Pareto's Rule and Its Extremal Set

It often happens that a certain rule with preference of DM $\preceq_d$ is reflexive but not connected. Since the preference on the extremal set[1] of X_A with $\preceq_d$, $M(X_A, \preceq_d)$ cannot be ordered for such a case, optimization on X_A is not well-defined. Hence if this is the case, the main concern is to obtain the whole extremal set $M(X_A, \preceq_d)$ or to introduce another rule by which a weak or strict order can be established on it.

The so-called Pareto optimal set[2] is defined as the extremal set of X_A with $\preceq_p$ such that the following Pareto's rule holds.

Pareto's rule: $x \succeq_p y \leftrightarrow x - y$ is contained in the nonnegative orthant.

Since the preference in terms of Pareto's rule is known to be only a partial order, it is impossible to order the preference on $M(X_A, \preceq_p)$ completely.

---

[1] If $\hat{x}$ is contained in the extremal set $M(X_A, \preceq_d)$, then there is no such x ($\neq \hat{x}$) that $x \succeq_d \hat{x}$ in X_A.

[2] The term non-inferior set or non-dominated set is used interchangeably.

Table C.1. Classification of multi-objective problems

	When	How to	Representative methods
			Representative methods
	Prior	Lottery	Optimize utility function
Optimi-zation	Gradual	Non-interactive inquiry	Optimal weighting method
			Lexicographical method
			Goal programming
		Interactive inquiry	Derived from single-objective optimization
			*Heuristic/Random search, IFW, SWT
			*Pair-wise comparison method, simplex method
			Interactive goal programming
			*STEM, RESTEM, Satisfying tradeoff method
	Preserved	Pair-wise comparison	AHP
			MOON^2, MOON^{2R}
Analysis	–	Schematic	ϵ-constraint method,
			weighting method, MOGA

However, noticing that $\preceq_p$ is a special case of $\preceq_d$ (this implies that $\preceq_p \subset \preceq_d$), the following relation will hold between extremal sets:

$$M(X_A, \preceq_p) \supset M(X_A, \preceq_d).$$

This implies that the Pareto optimality is the condition necessary at least in the multi-objective optimization. Hence, another rule becomes essential for choosing the preferentially optimal solution or the best compromise solution from the Pareto optimal set.

C.2 Traditional Methods

There are a variety of methods for MOP so far, and they are classified as summarized in Table C.1. Since the Pareto optimal solution plays an important role, its derivation has been a major interest in the earlier studies to the recent topics associated with metaheuristic approaches mentioned in Sect. 2.2. Roughly speaking, solution methods of MOP are classified into prior and interactive methods. Since the prior articulation methods try to reveal the preference of the DM prior to the search process, no articulation is done during the search process. On the other hand, the interactive methods can articulate the conflicting objectives adaptively and elaborately during the search process. For these reasons, the interactive methods are used popularly now.

C.2.1 Multi-objective Analysis

As mentioned already, obtaining the Pareto optimal solution (POS) set or non-inferior solution set is a primal procedure for MOP. Moreover, in the case where the number of objectives is small enough to depict POS set graphically,

say no more than three, it is possible to choose the best compromise solution
based on it. Therefore, a brief explanation of generating methods of the POS
set will be described below in the case where the feasible region is given by

$$X = \{x \mid g_i(x) \geq 0 \ \ (j = 1, \ldots, m), \ x \geq 0\}.$$

A. The Weighting Method and the ϵ-constraint Method

Both the weighting method and the ϵ-constraint method are well-known as
methods for generating the POS set. These methods are considered as the first
approaches to multi-objective optimization. According to the KKT conditions,
if $\hat{x}^*$ is a POS, then there exists such $w_j \geq 0$ and strictly positive for $\exists j$,
$(j = 1, \ldots, N)$ and $\lambda_j \geq 0$, $(j = 1, \ldots, m)$ that satisfy the following Pareto
optimal conditions[3]:

$$\begin{cases} \hat{x}^* \in X \\ \lambda_j g_j(\hat{x}^*) = 0 \quad (j = 1, \ldots, m) \\ \sum_{j=1}^{N} w_j (\partial f_j / \partial x)_{\hat{x}^*} - \sum_{j=1}^{m} \lambda_j (\partial g_j / \partial x)_{\hat{x}^*} = 0 \end{cases}.$$

Inferring from these conditions, we can derive the POS set by solving
the following single-objective optimization problem repeatedly while varying
weights of the objective functions parametrically [8]:

$$[Problem] \quad \min \ \sum_{j=1}^{N} w_j f_j(x) \quad \text{subject to} \quad x \in X.$$

On the other hand, the ϵ-constraint method is also formulated by the
following single-objective optimization problem:

$$[Problem] \quad \min \quad f_p(x)$$
$$\text{subject to} \quad \begin{cases} x \in X \\ f_j(x) \leq f_j^* + \epsilon_j \ (j = 1, \ldots, N, \ j \neq p) \end{cases},$$

where f_p and f_j^* represents a principal objective and an optimal value of
$f_j(x)$, respectively. Moreover, $\epsilon_j (> 0)$ is an amount of degradation of the j-th
objective function. In this case, by varying ϵ_j parametrically, we can obtain
the POS set.

From a computational aspect, however, these generation methods unfortunately require much effort to draw the whole POS set. Such efforts expand
as rapidly as the increase in the number of objective functions. Hence, these
methods are amenable for dealing with cases with two or three objectives
where the tradeoff on the POS set can be observed visually. They are useful
for generating a POS as a candidate solution in the iterative search process.

---

[3] These conditions are necessary and when all $f_j(x)$ are convex functions and X is
a convex set, they become sufficient as well.

C.2.2 Prior Articulation Methods of MOP

This section shows a few methods that belong to the prior articulation methods in the earlier stage of the studies. A common idea in this class is to obtain a unified objective function first, and derive a final solution from the resulting single-objective function.

A. The Optimal Weight Method

The best-compromise solution must be located on the POS set that is tangent to the indifference curve. Here, the indifference curve is a solution set that belongs to the equivalence class of a preferentially indifferent set.

Noticing this fact, Marglin [9] and Major [10] have shown that the slope of the tangent plane at the best compromise is proportional to the weights that represent a relative importance among the objectives. Hence if these weights, called the optimal weight w^*, are known beforehand, the multi-objective optimization problem refers to a usual single-objective problem,

$$[Problem] \quad \min \ \sum_{j=1}^{N} w_j^* f_j(x) \quad \text{subject to} \quad x \in X.$$

However, in general, since it is almost impossible to know such an optimal weight *a priori*, iteration becomes necessary to improve the preference of solution. Starting with an initial set of weights, the DM must adjust the weights to articulate the conflicting objectives. The major difficulty in this approach is that the optimal weight should be inferred in the absence of any knowledge about the POS set.

B. Hierarchical Methods

Though the optimal weight is hardly known *a priori*, we might rank the order of importance among the multiple objectives more easily. If this is true, it is possible to take a simple procedure as follows [11, 12]. Since the multiple objectives are placed in order of the relative importance, the first step tries to optimize the objective with the highest priority[4],

$$[Problem] \quad \min \ f_1(x) \quad \text{subject to} \quad x \in X. \tag{C.1}$$

After this optimization, the second problem will be given under the objective with the next priority,

$$[Problem] \quad \min \ f_2(x) \quad \text{subject to} \quad \begin{cases} x \in X \\ f_1(x) \le f_1^* + \Delta f_1 \end{cases},$$

where f_1^* and $\Delta f_1 (> 0)$ represent, respectively, the optimal value of Problem C.1 and the maximum amount of degradation allowed to improve the rest.

---

[4] The suffix is supposed to be renumbered in the order of importance.

Continuing this procedure in turn, the final problem will be solved for the objective with the lowest priority as follows:

$[Problem]$ min $f_N(x)$

subject to $\begin{cases} x \in X \\ f_j(x) \leq f_j^* + \Delta f_j \ (j = 1, \ \ldots, \ N-1) \end{cases}$.

Though the above procedures are intelligible, applications seem to be restricted mainly due to the two defects. It is often hard to order the objectives lexicographically following the importance beforehand. How to decide the allowable degradation in turn $(\Delta f_1, \Delta f_2, \ldots, \Delta f_{N-1})$ is another difficulty.

Consequently, these methods developed in the earlier stage seem to be applicable only to the particular situation in reality.

C. Goal Programming and Utility Function Theory

Goal programming was originally studied by Charnes and Cooper [13] for linear systems. Then it was extended and applied to many cases by many authors. A basic idea of the method relies on minimizing a weighted sum of the absolute deviations from an ideal goal,

$[Problem]$ min $\displaystyle\sum_{j=1}^{N} w_j |d_j|$

subject to $\begin{cases} x \in X \\ f_j(x) - \tilde{f}_j^* \leq d_j \ (j = 1, \ \ldots, \ N) \end{cases}$,

where $\tilde{f}_j^*$ is the ideal value for the j-th objective that is set forth by the DM, and each weight w_j should be specified according to the priority of the objective.

Goal programming has computational advantages particularly for linear systems with linear objective functions, since it refers to LP. In any case, it has a potential use when the ideal goal and weights can reflect the DM's preference precisely. It is quite difficult, however, to obtain such quantities without any knowledge about what tradeoffs are embedded in the POS set. In addition to it, it should be noticed that the improper selection of the ideal goal cannot yield a POS from this optimization. Therefore, setting the ideal goal is especially important for goal programming.

Utility function theory has been studied mainly in the field of economics and applied to some optimizations in engineering field. The major concerns of the method refer to the assessment of the utility function and its evaluation. The utility function is generally a function of multiple attributes that takes a greater value for the consequence more preferable to DM. The existence of such function is proven as shown in Theorem 2 in the preceding section.

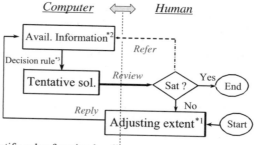

Identify value function locally
*1 Aspiration level (upper, lower), Marginal substitution rate,
    Trade-off interval
*2 Pay-off matrix (Utopia, Nadir), Sensitivity, Tradeoff curve
*3 Minimize distance/surrogate value, Pair-comparison

Fig. C.1. General framework of the solution procedure of an interactive method

Hence, if the explicit form of the utility function is known, MOP also refers to a single-objective optimization problem searching the alternative that possesses the highest utility in the feasible region. However, no general specification rules deciding a form of the utility function exist except for the condition that it must monotonically increase as the preference of the DM increases.

Identification of the utility function is, therefore, not an easy task and is peculiar to the problem under consideration. Since a simple form of the utility function is favorable for application, many efforts have been paid to obtain the utility function with a suitable form under mild conditions. The simplest additive form is derived under the conditions of the utility independence of each objective and the preference independence between the objectives. A detailed explanation regarding the utility function theory is found in other literatures [14, 6].

C.2.3 Some Interactive Methods of MOP

This class of methods relies on iterative procedures, each of which consists of a computational phase by computer and a judgment phase by DM. Through such human–machine interaction, the DM's preference is articulated progressively. Referring to the general framework depicted in Figure C.1, it is possible to invent many methods by combining reference items for adjusting, available information in tradeoff, and decision rules to obtain a tentative solution. Commonly, the DM is required to assess his/her preference based on the local information around a tentative solution or by direct comparison between the candidate solutions. Some of these methods will be outlined below.

Through the assessment of preferences in objective space, the Frank–Wolf algorithm of SOP is extended to MOP [15] assuming the existence of an aggregating preference function $U(f(x))$. $U(\cdot)$ is a monotonically increasing function with f, and is known only implicitly. The hill climbing technique employed in non-linear programming is used to increase the aggregating preference function most rapidly. For this purpose, the direction search problem is solved first through the value assessment of the DM to the tentative solution $\hat{x}^k$ at the k-th step,

$$[Problem] \quad \max \quad \sum_{j=1}^{N} w_j^k (-\partial f_j / \partial x)_{\hat{x}^k} y \quad \text{subject to} \quad y \in X,$$

where $w_j^k (j = 1, \dots, N)$ is defined as

$$w_j^k = (\partial U / \partial f_j) / (\partial U / \partial f_p)_{\hat{x}^k} \ (j = 1, \ \dots, \ N, \ j \neq p).$$

Since the explicit form of the aggregating function $U(f(x))$ is unknown *a priori*, the approximate values of w_j^k must be induced from the DM as the marginal rates of substitution (MRS) of each objective to the principal objective function f_p. Here MRS between f_p and f_j is defined as a rate of loss in f_p to the gain at $f_j, (j = 1, \dots, N, j \neq p)$ when the DM is indifferent to such changes while all other objectives are kept at their current values.

Then, a one-dimensional search is carried out in the steepest direction thus decided, *i.e.*, $y - \hat{x}^k$. By assessing the objective values directly, the DM is required to judge how far the most preferable solution will be located in that direction. The result provides an updated solution. Then going back to the direction search problem, the same procedures will be repeated until the best compromise solution is attained.

The defects of this method are as follows:

1. Correct estimation of the MRS is not easy in many cases, though it might greatly influence the convergence of the algorithm.
2. No significant knowledge about trade-off among the candidate solutions can be conceived by the DM, since most of the solutions obtained in the course of the iteration do not belong to the POS set.

In the method by Umeda *et al.* [16], the weighted sum of each objective function is used as a basis for generating a candidate solution,

$$[Problem] \quad \min \quad \sum_{j=1}^{N} w_j f_j(x) \quad \text{subject to} \quad \begin{cases} x \in X \\ \sum_{j=1}^{N} w_j = 1 \end{cases}.$$

Supposing that the candidate solution can be generated corresponding to the different sets of weights, the search incorporated with value assessment by the DM can be carried out conveniently in the parametric space of weights. The simplex method [17] in non-linear programming is used to search the optimal weights with a technique of pair-wise comparison for evaluating the preference between the candidates. The ordering among the vertices shown

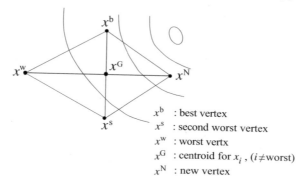

x^b : best vertex

x^s : second worst vertex

x^w : worst vertx

x^G : centroid for x_i , ($i\neq$worst)

x^N : new vertex

Fig. C.2. Solution process of the interactive simplex method

in Figure C.2 is carried out on the basis of preference instead of the values in the original SOP method. Since this method requires no quantitative reply from the DM, it seems suitable for the nature of human beings. However, the pair-wise comparison becomes increasingly troublesome and is likely to be inconsistent as the number of objective functions increases. It is possible to develop a similar algorithm in ϵ-space by using the ϵ-constraint method to derive a series of candidate solutions.

Geometrical understanding of MOP claims that the best compromise solution must be located at the point where the trade-off surface and the indifference surface are tangent with each other. Mathematically this requires that the tradeoff ratio to the principal objective is equivalent to the MRS at the best compromise point $\hat{f}^*$,

$$\beta_{pj}(\hat{f}^*) = m_{pj}(\hat{f}^*) \quad (j = 1, \ldots, N, \ j \neq p), \tag{C.2}$$

where β_{pj} and m_{pj} are the tradeoff ratio and the MRS of the j-th objective to the p-th objective, respectively. Noticing this fact, Haimes and Hall [18, 19] developed a method termed the surrogate worth tradeoff (SWT) method. In SWT, the tradeoff ratio can be obtained from the Lagrange multipliers for the active ϵ-constraint whose Lagrange function is given as follows:

$$L(x, \ \lambda) = f_p(x) + \sum_{j=1, j\neq p}^{N} \lambda_{pj}(f_j(x) - f_j^* - \epsilon_j),$$

where $\lambda_{pj}, (j = 1, \ldots, N, j \neq p)$ are Lagrange multipliers.

To express λ_{pj} or β_{pj} as a function of $f_p(x)$, Haimes and Hall used regression analysis. For this purpose, the ϵ-constraint problem is solved repeatedly by varying a certain $\epsilon_j, (\exists j \neq p)$ parametrically while keeping other ϵ constant. Instead of evaluating Equation C.2 directly, the surrogate worth function $W_{pj}(f)$ is introduced to reduce the DM's difficulties to work with this. The surrogate worth function is defined as a function that indicates the degree

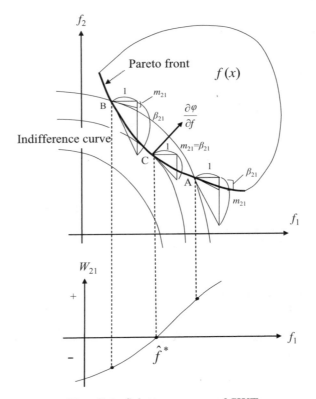

Fig. C.3. Solution process of SWT

of satisfaction of each objective with the specified objective in the candidate solution. This is usually an integer-valued function of ordinal scale varying on the interval $[-10, 10]$. The positive value of this function means that further improvement of the j-th objective is preferable as compared with the p-th objective, while the negative value corresponds to the opposite case. Therefore, the indifference band of the j-th objective is attained at the point where W_{pj} becomes zero, as shown in Figure C.3. Here, the indifference band is defined as a subset of the POS set where the improvement of one objective function is equivalent to the degradation of the other. In the SWT method, a technique of interpolation is recommended to decide this indifference band.

Based on the DM's assessment by the surrogate worth function, the best compromise solution will be obtained from the common indifference band of every objective. This is equivalent that the following conditions are satisfied:

$$W_{pj}(\hat{f}^*) = 0 \quad (j = 1, \ldots, N, j \neq p).$$

The major difficulty of this method is the computational load when assessing the surrogate worth function that expands rapidly as the number of

$$\begin{bmatrix} f_1^* & f_2(x_1^*) & \cdots & \cdots & f_N(x_1^*) \\ f_1(x_2^*) & f_2^* & \cdots & \cdots & f_N(x_2^*) \\ \vdots & \vdots & & & \vdots \\ \vdots & \vdots & & & \vdots \\ f_1(x_N^*) & f_2(x_N^*) & \cdots & \cdots & f_N^* \end{bmatrix}$$

Fig. C.4. Example of a Pay-off matrix

objectives increases. Additionally, the method has such a misunderstanding that the ordinal scale of the surrogate worth function is treated as if it might be cardinal.

The step method (STEM) developed by Benayoun *et al.* [20] is viewed as an interactive goal programming. In STEM, closeness to the ideal goal is measured by Minkowski's p-metric in objective space. ($p = \infty$ is chosen in their method.) At each step, the DM interacts with the computer to articulate the deviations from the ideal goal or to rank the relative importance under the multiple objectives. At the beginning of the procedure, a pay-off matrix is constructed by solving the following scalar problem:

$$[Problem] \quad \min \quad f_j(x) \quad \text{subject to} \quad x \in D^k \quad (\forall j \in I_u^{k-1})^5,$$

where D^k denotes a feasible region at the k-th step. It is set at the original feasible region initially, *i.e.*, $D^1 = X$.

The (i, j) element of the pay-off matrix shown in Figure C.4 represents the value of the j-th objective function evaluated by the optimal solution of the i-th problem x_i^*, *i.e.*, $f_j(x_i^*)$. This pay-off matrix provides helpful information to support the interactive correspondences. For example, a diagonal set of the matrix can be used to set up an ideal goal where any feasible solution cannot attain in any way. On the other hand, from a set of values in each column, we can observe the degree of variation or sensitivity of the objective with respect to the different solution, *i.e.*, $x_i^*, (i = 1, \ldots, N)$.

Since the preference will be raised by approaching the ideal goal, a solution nearest to the ideal goal may be chosen as a promising preferential solution. This idea leads to the following optimization problem, which is another form of the min-max strategy based on the L_∞ measurement in the generalized metric:

$$[Problem] \quad \min \quad \lambda \text{ subject to} \quad \begin{cases} x \in D^k \\ \lambda \geq w_j^k(f_j(x) - f_j^*) \ (j = 1, \ldots, N), \end{cases} \quad (C.3)$$

where w_j^k represents a weight on the deviation of the j-th objective value from its ideal value at the k-th step. It is given as $w_j^k = 1/f_j^*$ and $\sum_j w_j = 1$.

---

5 $I_u^0 = \{1, \ldots, N\}$

In reference to the pay-off matrix, the DM is required to classify each objective value of the resulting candidate solution $\hat{f}_j^k$ into a satisfactory class I_s^k and an unsatisfactory class I_u^k. Moreover, for $\forall j \in I_s^k$, the DM needs to respond the permissible amounts of degradation Δf_j that he/she can accept for the tradeoff. Based on these interactions, the feasible region is modified for the next step as follows:

$$D^{k+1} = D^k \cap \left\{ x \,\middle|\, \begin{array}{ll} f_j(x) \le \hat{f}_j^k + \Delta f_j & (^\forall j \in I_s^k) \\ f_j(x) \le \hat{f}_j^k & (^\forall j \in I_u^k) \end{array} \right\}. \tag{C.4}$$

Also, new weights are recalculated by setting the weights equal to zero for the objectives that have already been satisfied, $i.e.$, $\forall j \in I_s^k$. Then going back to Problem C.3, the same procedure will be repeated until the index set I_u^k becomes empty.

Shortcomings of this method are the following:

1. The ideal goal will not be updated along with the articulation. Hence the weights calculated based on the non-ideal values at the current step are likely to be biased.
2. Nevertheless it is not necessarily easy for the DM to respond the amounts of degradation Δf_j; the performance of the algorithm depends greatly on their proper selection.

The revised method of STEM termed RESTEM [21] has much more flexibility in the selection of degradation amounts, and also gives more information to aid the DMs interaction. This is brought about by updating the ideal goal at each step and by introducing a parameter that scales the weight properly. This method solves the following min-max optimization[6] to derive a candidate solution in each step:

$$[Problem] \quad \min \ \lambda \ \text{subject to} \ \begin{cases} x \in D^k \\ \lambda \ge w_j^k(f_j(x) - f_j^{*k}) \ (j = 1, \dots, N) \end{cases},$$

where f_i^{*k} denotes the ideal goal updated at each iteration given as follows:

$$f_i^{*k} = \{G_i^k, \ (^\forall i \in I_u^{k-1}), \ \hat{f}_i^{k-1}, \ (^\forall i \in I_s^{k-1})\},$$

where, G_i^k $(\forall i \in I_u^{k-1})$ denotes the i-th diagonal value of the k-th cycle pay-off matrix, and $\hat{f}_i^{k-1}$, $(\forall i \in I_s^{k-1})$ the preferential value at the preceding cycle.

---

[6] The following augmented objective function is amenable to obtaining practically the strict Pareto optimal solution:

$$[Problem] \quad \min \ \lambda + \epsilon \Big(\sum_{i \in I_u^{k-1}} w_i^k(f_i(x) - f_i^{*k}) + \sum_{i \in I_s^{k-1}} w_i^k(f_i(x) - \hat{f}_i^{k-1}) \Big),$$

where ϵ is a very small value.

Moreover, each weight w_i^k is computed by the following equation:

$$w_i^k = \alpha_i^k / \sum_{j=1}^{N} \alpha_j^k,$$

$$\text{where} \quad \alpha_i^k = \begin{cases} (1-\mu) \cdot \left(\dfrac{G_j^k - \hat{f}_j^{k-1}}{\hat{f}_j^{k-1}} \right) \left(\dfrac{1}{\hat{f}_j^{k-1}} \right), & (^\forall j \in I_u^{k-1}) \\[2ex] \mu \cdot \left(\dfrac{\Delta f_j^{k-1}}{\hat{f}_j^{k-1}} \right) \left(\dfrac{1}{\hat{f}_j^{k-1}} \right), & (^\forall j \in I_s^{k-1}) \end{cases},$$

where parameter μ is a constant to scale the degree of the DM's tradeoff between the objectives in I_s and I_u. When $\mu = 0$, the DM will try to improve the unsatisfied objectives at the expenses of the satisfied objectives by degrading by Δf_j^{k-1} in the next stage. This corresponds to the algorithm of STEM in which the selection of Δf_j^{k-1} plays a very important role. On the contrary, when $\mu = 1$, the preferential solution will stay at the previous one without taking part in the tradeoffs at all. By selecting a value between these two extremes, the algorithm can possess a flexibility against the improper selection of Δf_j. This property is especially important since every DM may not always conceive his/her own preference definitely.

Then the admissible region is revised as Equation C.4, and the same procedure will be repeated until every objective has been satisfied.

This method is successfully applied to a production system [22] and a radioactive waste management [23] system and its expansion planning [24]. Another method [25] uses another reference such as aspiration level to specify the preference region more compactly, and is more likely to lead the solution to the preferential optimum.

Evaluation of the interactive method was compared among STEM, IFW and a simple trial and error procedure [26]. A benchmark problem is solved on a fictitious company management problem under three conflicting objectives. Then the performance of the methods is evaluated by the seven measures listed below.

1. The DM's confidence in the best compromise solution.
2. Easiness of the method.
3. Understandability of the method logic.
4. Usefulness of information provided to aid the DM.
5. Rapidity of convergence.
6. CPU time.
7. Distance of best compromise solution from the efficient (non-inferior) surface.

Since the performance of the method is strongly dependent on the problem and the characteristics of the DM, no methods outperformed the others in all the above aspects.

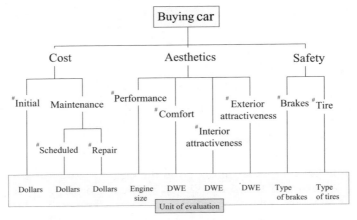

DWE: Direct Worth Estimate
(# : Leaf node)

Fig. C.5. Example of car selection

C.3 Worth Assessment and the Analytic Hierarchical Process

The methods described here enable us to make a decision under multi-objectives among a number of alternatives in a systematic and plain manner. We can use the methods for planning, setting priorities and selecting the best choice.

C.3.1 Worth Assessment

According to the concept of worth assessment [27, 28], an overall preference relation is described by the multi-attributed consequences or objectives that are structured in a hierarchy. In the worth assessment, the worth of each alternative is measured by an overall worth score into which every score should be combined. The worth score assigned to all possible values of a given performance measure must range commonly on the interval [0, 1]. This also provides a rather simple procedure to find out the best choice among a set of alternatives by evaluating the overall worth score.

Below, major steps of the worth assessment are shown and some explanations are given for an illustrative example regarding the best car selection as shown in Figure C.5.

Step 1: Place a final objective for the problem-solving under consideration at the highest level. (The "best" car to buy.)

Step 2: Construct an objective tree by dividing the higher level objectives into several lower level objectives in turn until the overall objectives can be defined in enough detail. ("Best" for the car selection is judged from three lower level indicators, *i.e.*, "cost, aesthetics, and safety". At the next step, "cost" is divided into "initial" and "maintenance", and so on.)

Step 3: Select an appropriate performance measure for each of the lowest level objectives. (Say, the initial cost in money (dollars).)

Step 4: Define a mathematical rule to assign a worth score to each value of the performance measure.

Step 5: Assign weights to represent a relative importance among the objectives that are subordinate to the same objective just by one level higher. (Child indicators that influence their parent indicator.)

Step 6: Compute an effective weight μ_i for each of the lowest level objectives (leaf indicators). This will be done by multiplying the weights along the path from the bottom to the top in the hierarchy.

Step 7: The effective weight is multiplied by the adjustment factor α_i that reflects the DM's confidence placed in the performance measures.

Step 8: Evaluate an overall worth score by $\sum_i \xi_i S_i(A_j)$, where $S_i(A_j)$ denotes the worth score of alternative A_j from the i-th performance measure and ξ_i an adjusted weight, *i.e.*, $\xi_i = \alpha_i \mu_i / \Sigma_i \alpha_i \mu_i$.

Step 9: Select the alternative with the highest overall worth score.

C.3.2 The Analytic Hierarchy Process (AHP)

The analytic hierarchy process (AHP) [29] is a multi-objective optimization method based on a hierarchy that structures the value system of the DM. By just carrying out the simple subjective judgments in terms of a pairwise comparison between decision elements, the DM can choose the most preferred solution among a finite number of decision alternatives. Just like the worth assessment method, it begins with constructing an objective tree through breaking down successively the upper level goals into their respective sub-goals[7] until a value system of the problem has been clearly defined. The top level of the objective tree represents a final goal relevant for the present problem-solving, while the decision alternatives are placed at the bottom level. The alternatives are connected to every sub-goal at the lowest level of the constructed objective tree. This last procedure is definitely different from the worth assessment method where the alternatives are not placed (see Figure C.6).

Then the preference data collected from the pair-wise comparisons mentioned below is used to compute a weight vector to represent a relative importance among the sub-goals. Though the worth assessment asks the DM directly respond to such weights, the AHP requires only the relative judgment through pair-wise comparison, which is easier for the DM. This is also different from the worth assessment method and a great advantage over it.

---

[7] It does not matter even if they are qualitative sub-goals like the worth assessment method.

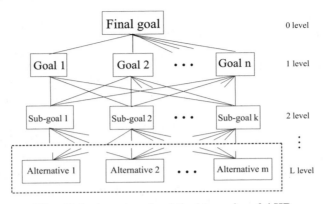

Fig. C.6. An example of the hierarchy of AHP

Table C.2. Conversion table.

Linguistic statement	a_{ij}
Equally	1
Moderately	3
Strongly	5
Very strongly	7
Extremely	9
Intermediate judgments	2,4,6,8

Finally, by using the aggregating weights over the hierarchy, the rating of each alternative is carried out to make a final decision.

At the data gathering step of AHP, the DM is asked to express his/her relative preference for a pair of sub-goals. Such responses take place by using linguistic statements, and are then transformed into the numeric score through the conversion table as shown in Table C.2. After doing such pair-wise comparisons repeatedly, a pair-wise comparison matrix A is obtained, whose i-j element a_{ij} represents a degree of relative importance for the j-th sub-goal f^j to the i-th f^i. Assuming that the value represents the rate of degree between the pair, $i.e.$, $a_{ij} = w_i/w_j$, we can derive two apparent relations like $a_{ii} = 1$ and $a_{ji} = 1/a_{ij}$. This means that we need only $N(N-1)/2$ pair-wise comparisons over N sub-goals. Moreover, transitivity in relation, $i.e.$, $a_{ij} \cdot a_{jk} = a_{ik}, (\forall i, j, k)$ must hold from the definition of the pair-wise comparison. Therefore, for example, if you say "I like apples more than oranges", "I like oranges more than bananas", and "I like bananas more than apples", you would be very inconsistent in your pair-wise judgments.

Eventually, the weight vector is derived from the eigenvector corresponding to the maximum eigenvalue λ_{max} of A. Equation C.5 is the eigenequation to calculate the eigenvector $\hat{w}$, which is normalized to be $\sum w_i = 1$[8],

$$(A - \lambda I)\hat{w} = 0, \tag{C.5}$$

where I denotes a unit matrix, and $w_i = \hat{w}_i(\lambda_{\max})/\sum_{i=1}^{N} \hat{w}_i(\lambda_{\max})$, $(i = 1, \ldots, N)$.

In practice, before computing the weights, a degree of inconsistency is measured by the consistency index CI defined by Equation C.6,

$$CI = \frac{\lambda_{\max} - N}{N - 1}. \tag{C.6}$$

Perfect consistency implies a value of zero of CI. However, perfect consistency cannot be demanded since subjective judgment of human beings is often biased and inconsistent. It is empirically known that the result is acceptable if $CI \leq 0.1$. Otherwise the pair-wise comparison should be revised before the weights are computed. There are several methods to fix various shortcomings associated with the inconsistent pair-wise comparisons as mentioned in Sect. 3.3.3.

Thus calculated weights for every cluster of the tree are used to derive the aggregating weights for the lowest level objectives that are directly connected to the decision alternatives. By adding the evaluation among the alternatives per each objective[9], the rating of the decision alternatives is completed from the sum of weighted evaluation since the alternatives are connected to all of the lowest level objectives. The largest rating represents the best choice. This totaling method is just the same as that of the worth assessment method.

The outstanding advantages of AHP are summarized as follows.

1. It needs only simple subjective judgments in the value assessment.
2. It is one of the few methods where it is possible to perform multi-objective optimization with both qualitative and quantitative attributes without paying any special attention.

These are the major reasons why AHP has been applied to various real world problems in many fields. In contrast, the great number of pair-wise comparisons necessary to do in the complicated applications is the inconvenience of AHP.

---

[8] There are some mathematical techniques such as eigenvalue, mean transformation, and row geometric mean methods.

[9] Just the same way as the weighting of the sub-goals is applied among the set of alternatives.

References

1. Wierzbicki AP, Makowski M, Wessels J (2000) Model-based decision support methodology with environmental applications. Kluwer, Dordrecht
2. Sen P, Yang JB (1998) Multiple criteria decision support in engineering design. Springer, New York
3. Osyczka A (1984) Multicriterion optimization in engineering with FORTRAN programs. Eliss Horwood, West Sussex
4. Zeleny M (1982) Multiple criteria decision making. McGraw-Hill, New York
5. Cohon JL (1978) Multiobjective programming and planning. Academic Press, New York
6. Keeney RL, Raiffa H (1976) Decisions with multiple objectives: preferences and value tradeoffs. Wiley, New York
7. Lasdon LS (1970) Optimization theory for large systems. Macmillan, New York
8. Gass S, Saaty T (1955) The computational algorithm for the parametric objective function. Naval Research Logistics Quarterly, 2:39–45
9. Marglin SA (1967) Public investment criteria. MIT Press, Cambridge
10. Major DC (1969) Benefit-cost ratios for projects in multiple objective investment programs. Water Resource Research, 5:1174–1178
11. Benayoun R, Tergny J, Keuneman D (1970) Mathematical programming with multi-objective functions: a solution by P. O. P., Metra, 9:279–299
12. van Delft A, Nijkamp P (1977) The use of hierarchical optimization criteria in regional planning. Journal of Regional Science, 17:195–205
13. Charnes A, Cooper WW (1977) Goal programming and multiple objecive optimizations–part 1. European Journal of Operational Research, 1:39–54
14. Fishburn PC (1970) Utility theory for decision making. Wiley, New York
15. Geoffrion AM (1972) An interactive approach for multi-criterion optimization with an application to the operation of an academic department. Management Science, 19:357–368
16. Umeda T, Kobayashi S, Ichikawa A (1980) Interactive solution to multiple criteria problems in chemical process design. Computer & Chemical Engineering, 4:157–165
17. Nelder JA, Mead R (1965) Simplex method for functional minimization. Computer Journal, 7:308–313
18. Haimes YY, Hall WA (1974) Multiobjectives in water resource systems analysis: the surrogate worth trade off method. Water Resource Research, 10:615–624
19. Haimes YY (1977) Hierarchical analyses of water resources systems: modeling and optimization of large-scale systems. McGraw-Hill, New York
20. Benayoun R, Montgolfier de J, Tergny J (1971) Linear programming with multiple objective functions: step method (STEM). Mathematical Programming, 1:366–375
21. Takamatsu T, Shimizu Y (1981) An interactive method for multiobjective linear programming (RESTEM). System and Control, 25:307–315 (in Japanese)
22. Shimizu Y, Takamatsu T (1983) Redesign procedure for production planning by application of multiobjective linear programming. System and Control, 27:278–285 (in Japanese)
23. Shimizu Y (1981) Optimization of radioactive waste management system by application of multiobjective linear programming. Journal of Nuclear Science and Technology, 18:773–784

24. Shimizu Y (1983) Multiobjective optimization for expansion planning of rad-waste management system. Journal of Nuclear Science and Technology, 20:781–783

25. Nakayama H (1995) Aspiration level approach to interactive multi-objective programming and its applications. In: Pardolas PM et al.(eds.)Advances in Multicriteria Analysis, Kluwer, pp. 147-174

26. Wallenius J (1975) Comparative evaluation of some interactive approach to multicriterion optimization. Management Science, 21:1387–1396

27. Miller JR (1967) A systematic procedure for assessing the worth of complex alternatives. Mitre Co., Bedford, MA., Contract AF 19, 628:5165

28. Farris DR, Sage AP (1974) Worth assessment in large scale systems. Proc. Milwaukee Symposium on Automatic Controls, pp. 274–279

29. Saaty TL (1980) The analytic hierarchy process. McGraw-Hill, New York

Appendix D

The Basis of Neural Networks

In what follows, the neural networks employed for the value function modeling in Sect. 3.3.1 are introduced briefly, while leaving the detailed description to another book [1]. Another type known as the cellular neural network appeared in Chap. 4 for intelligent sensing and diagnosis problems.

D.1 The Back Propagation Network

The back propagation (BP) network [5, 2] is a popularly known feedforward neural network as depicted in Figure D.1. It consists of at least three layers of neurons fully connected to those at the next layer. They are an input layer, middle layers (sometimes referred to hidden layers), and an output layer. The number of neurons and layers in the middle should be changed based on the complexity of problem and the size of inputs.

A randomized set of weights on the interconnections is used to present the initial pattern to the network. According to an input signal (stimulus), each neuron computes an output signal or activation in the following way. First,

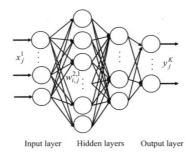

Input layer   Hidden layers   Output layer

Fig. D.1. A typical structure of the BP network

the total input x_j^n is computed by multiplying each output signal y_i^{n-1} times the random weight on that interconnection $w_{ij}^{n,n-1}$,

$$x_j^n = \sum_i w_{ij}^{n,n-1} y_i^{n-1}, \ \forall j \in n-- \text{ layer}.$$

Then this weighted sum is transformed by using an activation function $f(x)$ that determines the activity generated in the neuron by the input signal. A sigmoid function is typically used for such a function. It is a continuous, S-shaped and monotonically increasing function and asymptotically tends to the fixed value as the input approaches $\pm\infty$. Setting the upper limit to 1 and the lower limit to 0, the following formula is widely used for this transformation:

$$y_j^n = f(x_j^n) = 1/(1 + \exp^{-(x_j^n + \theta_j)}),$$

where θ_j is a threshold. Throughout the network, outputs are treated as inputs to the next layer. Thus the computed output at the output layer from the forward activation is compared with the desired target output values to modify the weights iteratively. The most widely used method of the BP network tries to minimize the total squared error in terms of the δ–rule. It starts with calculating the error gradient δ_j for each neuron on the output layer K,

$$\delta_j^K = y_j^K (1 - y_j^K)(d_j - y_j^K),$$

where d_j is the target value for output neuron j.

Thus the error gradient is determined for the middle layers by calculating the weighted sum of errors at the previous layer,

$$\delta_j^n = y_j^n (1 - y_j^n) \sum_k \delta_k^{n+1} w_{kj}^{n+1,n}.$$

Likewise, the errors are propagated backward one layer. The same procedure is applied recursively until the input layer has been reached. To update the network weights, these error gradients are used together with a momentum term that adjusts the effect of previous weight changes on present ones to adjust the convergence property,

$$w_{ij}^{n,n-1}(t+1) = w_{ij}^{n,n-1}(t) + \Delta w_{ij}^{n,n-1}(t)$$

and

$$\Delta w_{ij}^{n,n-1}(t) = \beta \delta_j^n y_i^{n-1} + \alpha \Delta w_{ij}^{n,n-1}(t-1),$$

where t denotes the iteration number, β the learning rate or the step size during the gradient descent search, and α a momentum coefficient, respectively.

In the discussion so far, the BP is viewed as a descent algorithm that tries to minimize the average squared error by moving down the contour of

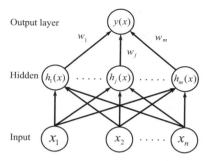

Output layer

Hidden

Input

Fig. D.2. Traditional structure of the RBF network

the error curve. In real world applications, since the error curve is a highly complex and multi-modal curve with various valleys and hills, training the network to find the lowest point becomes more difficult and challenging. The following are useful common training techniques [3]:

1. Reinitialize the weights: This can be achieved by randomly generating the initial set of weights each time the network is made to learn again.
2. Add step change to the weights: This can be achieved by varying each weight by adding about 10% of the range of the oscillating weights.
3. Avoid over-parameterization: Since too many neurons in the hidden layer cause poor predictions of the model, the network design with reasonable limits is desirable.
4. Change the momentum term: Experimenting with different levels of the momentum term will lead to the optimum very rapidly.
5. Avoid repeated or less noisy data: As easily estimated, duplicated information is harmful to generalizing their features. This can also be achieved by adding some noise to the training set.
6. Change the learning tolerance: If the learning tolerance is too small, the learning process never stops, while a too large tolerance will result in poor convergence. The tolerance level should be adjusted adequately so that no significant change in weights is observed.

D.2 The Radial-basis Function Network

The radial basis function (RBF) network [4] is another feedforward neural network whose simple structure (one output) is shown in Figure D.2. Each component of input vector x feeds forward to the neuron at the middle layer whose outputs are linearly combined with the weight w to derive the output,

$$y(x) = \sum_{j=1}^{m} w_j h_j(x),$$

where y denotes an output of the network and w a weight vector on the interconnection between the middle and output layers. Moreover, $h_j(\cdot)$ is an output from the neuron at the middle layer or input to the output layer.

The activation function of the RBF network is a radial basis function that is a special class of function whose response decreases (or increases) monotonically with distance from a center. Hence, the center, the distance scale, and the type of the radial function become key parameters of this network. A typical radial function is the Gauss function that is described, for simplicity, for a scalar input as

$$h(x) = \exp(-\frac{(x-c)^2}{r^2}),$$

where c denotes the center and r the radius.

Using a training data set such as (x^i, d^i), $(i = 1, \ldots, p)$, an accompanying form of the sum of the squared error E is minimized with respect to the weights (d^i denotes an observed output for input x^i),

$$E = \sum_{i=1}^{p}(d^i - y(x^i))^2 + \sum_{j=1}^{m}\lambda_j w_j^2, \tag{D.1}$$

where $\lambda_j, (j = 1, \ldots, m)$ denotes regularization parameters to prevent the individual data from sticking to too much or from overlearning. For a single hidden layer network with the activation function fixed in position and size, the expensive computation of the gradient descent algorithms used in the BP network is unnecessary for the training of the RBF network. The above least square problem refers to a simple solution of the m-dimensional simultaneous equations described in matrix form as follows:

$$Aw = H^T d,$$

where A is a variance matrix, and H a design matrix given by

$$H = \begin{bmatrix} h_1(x^1) & h_2(x^1) & \cdots & h_m(x^1) \\ h_1(x^2) & h_2(x^2) & \cdots & h_m(x^2) \\ \cdot & \cdot & \cdot & \cdot \\ \cdot & \cdot & \cdot & \cdot \\ \cdot & \cdot & \cdot & \cdot \\ h_1(x^p) & h_2(x^p) & \cdots & h_m(x^p) \end{bmatrix}.$$

Then A^{-1} is calculated as

$$A^{-1} = (H^T H + \Lambda)^{-1},$$

where Λ is a diagonal matrix whose elements are all zero except for those composed of the regularization parameters, $i.e.$, $\{\Lambda\}_{ii} = \lambda_i$. Eventually, the optimal weight vector that minimizes Equation D.1 is given as

$$w = A^{-1}H^T y.$$

Favorably, the RBF network enables us to model the value function adaptively depending on the unsteady decision environment often encountered in real world problems. For example, in the case of adding a new training pattern $p+1$ after p, the update calculation is given by Equation D.4 using the relations in Equations D.2 and D.3,

$$A_p = H_p^T H_p + \Lambda, \tag{D.2}$$

$$H_{p+1} = \begin{bmatrix} H_p \\ h_{p+1}^T \end{bmatrix}, \tag{D.3}$$

$$A_{p+1}^{-1} = A_p^{-1} - \frac{A_p^{-1}h_{p+1}h_{p+1}^T A_p^{-1}}{1 + h_{p+1}^T A_p^{-1} h_{p+1}}, \tag{D.4}$$

where $H_p = (h_1, h_2, \ldots, h_p)$ denotes the design matrix of the p-pattern.

On the other hand, when removing an i-th old training pattern, we use the relation in Equation D.5,

$$A_{p-1}^{-1} = A_p^{-1} + \frac{A_p^{-1}h_i h_i^T A_p^{-1}}{1 + h_i^T A_p^{-1} h_i}. \tag{D.5}$$

Since the load required for these post-analysis operations[1] are considerably reduced, the effect of time saving is obvious as the problem size becomes large.

References

1. Wasserman (1989) Neural computing: theory and practice. Van Nostrand Reinhold, New York
2. Bhagat P (1990) An introduction to neural nets. Chemical Engineering Progress, 86:55–60
3. Chitra SP (1993) Use neural networks for problem solving. Chemical Engineering Progress, 89:44–52
4. Orr MJL (1996) Introduction to radial basis function networks. http://www.cns.uk/people/mark.html
5. Rumelhart DE, Hinton GE, Williams RJ (1986) Learning representations by back-propagating errors. Nature, 323:533–536

---

[1] Likewise, it is possible to provide increment/decrement operations regarding the neurons [4].

Appendix E

The Level Partition Algorithm of ISM

In what follows, the algorithm of ISM method [1, 2] will be explained by limiting the concern mainly to its level partition. This procedure starts with defining the binary relation R on set S composed of n elements $(s_1, s_2, \ldots, s_n)$. Then it is described as $s_i R s_j$ if s_i has relation R with s_j. The ISM is composed of the following three major steps.

1. Enumerate elements to be structured in S, $\{s_i\}$.
2. Describe a context or content of relation R to specify a pair of the elements.
3. Indicate a direction of the relation between every pair of element $s_i R s_j$.

Viewing each element and relation as node and edge, respectively, such a consequence can be represented by a digraph as shown in Figure E.1. For numerical processing, however, it is more convenient to describe it by a binary matrix whose (i, j) element is given by representing the following conditions:

$$\begin{cases} a_{ij} = 1, & \text{if } i \text{ relates with } j \\ a_{ij} = 0, & \text{otherwise} \end{cases}.$$

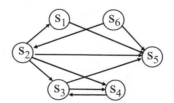

Fig. E.1. Example of a digraph

The collection of such a relationship over every pair builds a binary matrix. From the thus derived matrix A, called the adjacency matrix, the reachability matrix T is derived by repeating the following matrix calculation on the basis of Boolean algebra:

$$T = (A + I)^{k+1} = (A + I)^k \neq (A + I)^{k-1}.$$

Then, two kinds of set are defined as follows:

$$\begin{cases} R(s_i) = \{s_i \in S | m_{ij} = 1\} \\ A(s_i) = \{s_i \in S | m_{ji} = 1\} \end{cases},$$

where $R(s_i)$ and $A(s_i)$ denote a reachable set from s_i and an antecedent set to s_i, respectively. In the following, $R(s_i) \cap A(s_i)$ means the cap of $R(s_i)$ and $A(s_i)$.

Finally, the following procedure derives the topological relation or hierarchical relationship among the nodes (level partition):

Step 0: Let $L_0 = \phi$, $T_0 = T$, $S_0 = S$, $j = 1$.
Step 1: From T_{j-1} for S_{j-1}, obtain $R_{j-1}(s_i)$ and $A_{j-1}(s_i)$.
Step 2: Let us identify the element that holds $R_{j-1}(s_i) \cap A_{j-1}(s_i) = R_{j-1}(s_i)$, and include it in L_j.
Step 3: If $S_j = S_{j-1} - L_j = \{\phi\}$, then stop. Otherwise, let $j := j + 1$ and go back to Step 1.

The result of the level partition makes the set L group into its subset L_i as follows:

$$L = L_1 \cdot L_2 \cdot, \ldots, \cdot L_M,$$

where L_1 stands for the set whose elements belong to the top level, and L_M locates at the bottom level. Finally, ISM can reveal a topological configuration of the entire members of the system. For example, the foregoing graph is described as shown in Figure E.2.

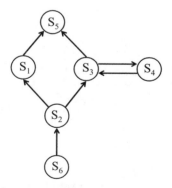

Fig. E.2. ISM structural model

Based on the above procedure, it is possible to identify the defects considered in the value function modeling in Sect.3.3.1. First let us recall that from the definition of the pair-wise comparison matrix (PWCM), any of the following relations holds:

- If $f^i \succ f^j$, then $a_{ij} > 1$.
- If $f^i \sim f^j$, then $a_{ij} = 1$.
- If $f^i \prec f^j$, then $a_{ij} < 1$.

Hence transforming each element of PWCM using the relation

- $a'_{ij} = 1$,   if $a_{ij} > 1$,
- $a'_{ij} = 1^*$,   if $a_{ij} = 1$,
- $a'_{ij} = 0$,   if $a_{ij} < 1$,

we can transform the PCWM into a quasi-binary matrix. Here, to deal with the indifference case $(a_{ij} = 1)$ properly in the level partition of ISM, notation 1^* is introduced, and defined by the following pseudo-Boolean algebra:

- $1 \times 1^* = 1^*,\ 1^* \times 1^* = 0,\ 1^* \times 0 = 0$
- $1^* + 1^* = 1,\ 1 + 1^* = 1,\ 1^* + 0 = 1^*$

Then, at each level L_k revealed by applying the level partition of ISM, we have the consequence where $R_{L_k}(s_i) \neq s_i, \forall s_i \in L_k$ causes a conflict on the transitivity. Here R_{L_k} denotes the reachable set from s_i in level L_k.

References

1. Sage AP (1977) Methodology for large-scale systems. McGraw-Hill, New York
2. Warfield JN (1976) Societal systems. Wiley, New York

Index

δ-rule, 298
ϵ-constraint method, 86, 280
ϵ-constraint problem, 92
0-1 program, 67
2-D CDWT, 189, 212
2-D DWT, 189

abnormal detection, 159
abnormal sound, 147
ACO, 34
activation function, 298, 300
activity modeling, 259
adaptive DE , see ADE
ADE, 30
adjacency matrix, 303
admissibility condition, 161, 173
age, 30
agent
  architecture, 230
  communication language, 230, 232
  definition of, 229
  matchmaking agent, 242
  performative, 230
  standard for, 230
aggregating function, 80
AGV, 2, 53
AHP, 89, 291
alignment, 32
alternative, 291
analytic hierarchy process, see AHP
annealing schedule, 24
ant colony algorithm, see ACO
AR model, 147

artificial variable, 265
aspiration criteria, 27
aspiration level, 289
associative memory, 7, 125, 155
automated guided vehicle, see AGV

B & B method, 268
back propagation, see BP
basic fast algorithm, 169
basic solution, 265
basic variable, 264
bi-orthogonal condition, 180, 183
bill of materials, see BOM
binary coding, 18
binary relation, 277, 303
binominal crossover, 28
boid, 32
BOM, 228, 230
Boolean algebra, 304
BP, 88, 113, 297
branch-and-bound method, see B & B
  method
building block hypothesis, 21

CDWT, 182
cellular cutomata, 126
cellular neural network, see CNN
changeover cost, 109
Chinese character pattern, 152
chromosome, 15, 62
CNC, 2, 53
CNN, 7, 11, 126, 128, 131, 155
coding, 15
  of DE, 28

cohesion, 32
COM, 226
combinatorial optimization, 14
complex discrete wavelet transform, *see* CDWT
complex method, 273
compromise solution, 6, 81, 92, 108
computerized numerical control, *see* CNC
consistency index, 90, 293
constraint, 3, 259
continuous wavelet transform, *see* CWT
contraction, 272
control, 3, 259
conversion table, 292
convex combination, 35
convex function, 270
convex set, 270
cooling schedule, 24
cooperation, 57
CORBA, 226
coupling constraint, 41
cross validation, 113
crossover, 15
    of ADE, 31
    of DE, 28, 29
crossover rate, 28, 63
    of ADE, 31
crowding distance, 84
CWT, 160
cycle time, 48

database, 223–226, 236
    integration, 225
DCOM, 226
DE, 27
decision maker, *see* DM
design matrix, 301
design of experiment, *see* DOE
diagnosis system, 7
differential evolution, *see* DE
digraph, 303
Dijkstra method, 45, 269
direct search, 274
discrete wavelet transform, *see* DWT
dispatching rule, 39, 58
distribution, 57
diversified generation, 35
DM, 79, 86, 88, 96, 98, 111, 121

virtual, 105, 109
DOE, 101
dual wavelet, 180
dual-tree algorithm, 182, 188
due time, 109
DWT, 180, 189, 205

EA, 2, 6, 14, 79
ECG, 209
eigenvalue, 90, 293
eigenvector, 293
elitism, 84
elitist preserving selection, 18
enhancement, 35
Enterprise Resource Planning, *see* ERP
Enterprise Systems, 221, 224
ERP, 221, 222, 224, 225, 237, 250
evolutionary algorithm, *see* EA
expansion, 272
expected-value selection, 17
exponential cooling schedule, 53
exponential crossover, 29
EXPRESS, 240
extreme condition, 270
extreme point, 265

fast algorithm, 160, 167, 180
feasibility, 265
FEM, 116
finite element method, *see* FEM
FIPA, 230–232, 235
fitness, 15, 62
flexibility, 60
flexibility analysis, 68
flow shop scheduling, 110
Foundation for Intelligent Physical Agents, *see* FIPA
Fourier transform, 8
fractal analysis, 194
Frank–Wolf algorithm, 283

GA, 14
Gantt chart, 55, 56, 110
Gauss function, 300
Gaussian function, 175
gene, 15
generalized path construction, 34
generation, 15
genetic algorithm, *see* GA

genetic operation, 81
genotype, 15
global
    best, 33
    optimization, 22, 269
    optimum, 6, 14, 21, 47
goal programming, 282
gradient method, 274
grain of quantization, 94
Gray coding, 21
greedy algorithm, 24

Hamming distance, 68, 95
Hamming distances, 140
Hannning window, 175
hard variable, 68
hierarchical method, 281
Hilbert pair, 176
Hopfield network, 128
hybrid approach, 7, 36
hybrid tabu search, 44, 45, 65, 69

ideal goal, 287
IDEF0, 3, 225, 259
idle time, 48, 49, 51, 57
ill-posed problem, 94
image processing, 7, 159
incommensurable, 77
increment operation, 117
indifference, 277
    band, 286
    curve, 91, 109, 281
    surface, 285
individual, 15
information technology, see IT
injection period, 49
injection sequencing, 39, 48
input, 3, 259
integer program, see IP
integrated optimization, 105
intelligent agent, 5
interactive method, 96, 279, 283
interior-point method, 268
interpolation, 185
Interpretive Structural Modeling, see
    ISM
inventory, 4, 66
IP, 37, 45, 268
ISM, 99, 303

ISO
    ISO 10303, 237, 240
    ISO 13584, 237
    ISO 15531, 237
    ISO 15926, 243, 244
    ISO 62264, 237, 239
    ISO TC184, 236
IT, 3, 5

JADE, 235
Java Theorem Prover, see JTP
job, 54, 55
job shop scheduling, 58
JRMI, 227
JTP, 235, 242, 243

Karush–Kuhn–Tucker condition, see
    KKT condition
KIF, 235, 242
KKT condition, 270, 280
knocking detection, 200
KQML, 230–232

Lagrange function, 270
Lagrange multiplier, 43, 270, 285
lead time, 4
learning rate, 298
least squares method, 195
level partition, 303
line stoppage, 39, 48, 51
linear programming, see LP
liver illness, 143
local
    best, 32
    optimum, 14, 17, 26
local optimum, 269
local search, 23, 26, 34
logistic, 38, 39, 65
long term memory, 27
lower aspiration level, 103
LP, 107, 264, 268

makespan, 108
Manufacture Resource Planning, see
    MRPII
Manufacturing Execution Systems, see
    MES
manufacturing system, 1, 222, 225
marginal rates of substitution, see MRS

master–slave configuration, 38
Material Requirements Planning, *see* MRP
mathematical programming, *see* MP
maximum entropy method, 147
Maxwell–Boltzmann distribution, 24
MCF, 44, 45, 269
mechanism, 3, 259
memetic algorithm, 34
merge, 56
merging, 81
MES, 224, 225
Message Oriented Middleware, *see* MOM
meta-model-base, 101
metaheuristic, 5, 6, 9, 13
MILP, 106
min-max strategy, 287
minimum cost flow problem, *see* MCF
MIP, 36, 108, 268
mixed-integer linear program, *see* MILP
mixed-integer program, *see* MIP
mixed-model assembly line, 38, 48
MMT-CNN, 140, 152, 153
MOEA, 79
MOGA, 82, 108
MOHybGA, 107
MOM, 227
momentum term, 298
MOON^2, 88, 96
MOON^{2R}, 88, 96
MOP, 5, 6, 9, 77, 277
MOSC, 96
mother wavelet, *see* MW
MP, 36, 263
MRA, 180
MRP, 221, 237, 250
MRPII, 221, 237
MRS, 284, 285
Multi-agent Systems, 229, 232, 242
multi-objective analysis, 86, 279
multi-objective evolutionary algorithm, *see* MOEA
multi-objective genetic algorithm, *see* MOGA
multi-objective optimization, *see* MOP
multi-objective scheduling, 105, 108
multi-resolution analysis, *see* MRA
multi-skilled operator, 54

multi-start algorithm, 21
multi-valued output function, 131
multiple allocation, 39
multiple memory tables, *see* MMT-CNN
mutant vector, 28, 29
  of ADE, 30
mutation, 15, 19
  of ADE, 30
  of DE, 28
mutation rate, 63
MW, 160, 161, 173

nadir, 88, 116
natural selection, 14
neighbor, 23, 26, 32, 53
neighborhood, 24, 126, 135
network linear programming, 268
neural network, *see* NN
neuron, 297
Newton–Raphson method, 274
niche count, 84
niche method, 82
niched Pareto genetic algorithm, *see* NPGA
NLP, 36, 269
NN, 2, 5, 6, 126, 297
non-basic variable, 264
non-dominance, 82
non-dominated rank, 85
non-dominated sorting genetic algorithm, *see* NSGA-II
non-inferior solution set, 279
nonlinear network, 127
nonlinear programming problem, *see* NLP
NP-hard, 41, 52, 69
NPGA, 84
NSGA, 108
NSGA-II, 84
numerical differentiation, 96

objective tree, 290
offspring, 18
one-point crossover, 94
ontology, 5, 9, 233, 235, 240, 241, 259
  languages, 240
  upper ontology, 243
OPC, 227
operation, 54, 55

optimal weight, 281
optimal weight method, 281
optimality, 265
orthogonal wavelet, 180
output, 3, 259
output function, 126
overlearning, 300
OWL, 240–242, 248

pair-wise comparison, 88, 89, 284, 291
pair-wise comparison matrix, see
    PWCM
parallel computing, 34, 38
parent, 18
Pareto
    domination tournament, 84
    front, 79
    optimal condition, 280
    optimal solution, 78, 279
    optimal solution set, see POS set
    ranking, 82
    rule, 278
Pareto-based, 80, 82
particle swarm optimization, see PSO
pay-off matrix, 287, 288
PDCA cycle, 5, 101
penalty coefficient, 107
penalty function, 37, 61, 267
permanently feasible region, 69
phenotype, 15
pheromone trail, 34
physical quantity, 246, 247
piecewise linear function, 127
pivot, 266
population, 15
population-based, 60, 79, 94
POS set, 78, 92, 108, 278–281
position, 32
positive definite, 270
positive semi-definite, 270
post-operation, 57
pre-operation, 57
preference relation, 277
preferentially optimal solution, 79, 102
premature convergence, 18
prior articulation method, 88, 279, 281
process, 54
process control, 221, 237
    systems, 221

production scheduling, 4
proportion selection, 17
PSO, 32
Publish and Subscribe, 227
PWCM, 89, 108, 292, 305

QP, 269
quadratic programming, see QP

radial basis function, see RBF
ranking selection, 17
RBF, 88, 299, 300
reachability matrix, 304
real number coding, 22, 27, 32
real signal mother wavelet, see RMW
reference point, 91
reference set, 34
reflection, 271
regularization parameter, 300
Remote Procedure Call, see RPC
reproduction, 15, 16
resource, 1, 8, 54, 55
response surface method, 101
RESTEM, 288
revised simplex method, 117
RI-spline wavelet, 160, 162, 182, 206
RMW, 162, 174
Rosenbrock function, 31
roulette selection, 17
RPC, 226

SA, 22, 39, 52, 110
saddle point, 270
SC, 5, 6, 77, 87
scaling
    function, 181, 183
scaling technique, 16
scatter search, 34
scheduling problem, 39, 54
schemata, 21
SCM, 38, 39, 65
selection, 81
    of ADE, 31
    of DE, 27, 29
self-similarity, 194
separation, 32
sequential quadratic programming, see
    SQP
service level, 66

shared fitness, 84
sharing function, 82, 95
short term memory, 26
short time Fourier transform, 159
shortest path problem, 45, 269
shuffling, 81
sigmoid function, 298
signal analysis, 7, 159
signal processing, 5
Simple Object Access Protocol, *see*
    SOAP
simplex method, 265, 271, 284
simplex tableau, 266
simulated annealing, *see* SA
simulation-based, 101, 116
single allocation, 39
singular value decomposition, *see* SVD
slack variable, 265
small-lot-multi-kinds production, 38
SOAP, 228, 239
soft computing, *see* SC
soft variable, 68
speciation, 81
spline wavelet, 162, 185
SQP, 113, 275
stable equilibrium point, 131
standard form, 264, 265
stationary condition, 16, 270
steady signal, 7
STEM, 287
step method, *see* STEM
stochastic optimization, 60
strict Pareto optimal solution, 288
strict preference, 277
string, 15
subjective judgment, 79, 104
supply chain management, *see* SCM
surrogate worth function, 285
surrogate worth tradeoff method, *see*
    SWT
SVD, 130
sweep out, 267
SWT, 285
symmetric property, 162
systems thinking, 1

tabu list, 26, 46
tabu search, *see* TS
tabu tenure, 26

tabu-active, 27
target vector, 28
temperature, 23
TI de-noising, 205
time-frequency analysis, 159
time-frequency method, 8
tournament selection, 18
tradeoff
    ratio, 285
    surface, 285
tradeoff analysis, 69, 79, 92
training data, 90
transition probability, 24
transitivity, 292
tri-valued output function, 143
trial solution, 88, 89
trial vector, 29
TS, 26, 44, 46
two-phase method, 267

uncertainty, 6, 60, 66
unconstrained optimization, 269
unsteady signal, 7
upper aspiration level, 103
utility function, 105, 278
utility function theory, 282
utopia, 88, 116

value function, 78
vector evaluated genetic algorithm, *see*
    VEGA
VEGA, 80
velocity, 32

wavelet, 5
    instantaneous correlation, *see* WIC
    scale method, *see* WSE
    shrinkage, 205, 214
    transform, 8, 11, 159
Web Services, 227, 228
weighting method, 280
WIC, 162, 203
Wigner distribution, 8, 159
window function, 174
WIP, 48, 53
work-in-process, *see* WIP
worth assessment, 290
WSM, 197

XML, 227, 239, 241